POISONED HARVEST

A Consumer's Guide to Pesticide Use and Abuse

Christopher Robbins

D1078520

VICTOR GOLLANCZ LTD
LONDON   1991

First published in Great Britain 1991
by Victor Gollancz Ltd
14 Henrietta Street, London WC2E 8QJ

A Gollancz Paperback Original

A CIP catalogue record for this book is available
from the British Library.

ISBN 0 575 04797 6

Photoset in Great Britain by
Rowland Phototypesetting Ltd, Bury St Edmunds, Suffolk.
Printed and bound by Cox & Wyman Ltd, Reading, Berks.

To the memory of my father
Dr Lincoln Robbins

Contents

Acknowledgements

Such a complicated subject could not have been unravelled without the significant help of supportive individuals and organisations. Help in the form of printed information, long telephone chats, or hours of recorded interviews was given by Elizabeth Bradbury of Waitrose; Tony Combes of Safeway; Richard Pugh of Tesco; John Love, Ross McLaren, Roy Spencer, and Marian Archer of Sainsbury's; Tom Clayton and Viv Jawett at Marks & Spencer; Bill Shannon, Keith Davies, and Martin Henderson at the Co-operative Wholesale Society; Derek Reid of Premier Brands; Colin Walker at the Institute of Terrestrial Ecology; Alastair Burn of the National Farmers' Union; Mr Ambrose of the Health and Safety Executive; Virginia Murray of the National Poisons Unit in Guy's Hospital, London; Peter Hurst of the Pesticides Trust; helpful staff at the MAFF; Martin Hegmann, a glasshouse grower and biological control enthusiast in Essex; Robert Pickford of Humber Growers, who has developed friendly insecticides for glasshouses; and Jackie Gear of the Henry Double-day Research Association. I am grateful to all the food and farming companies and organisations who responded to my long and detailed questionnaire which provided such valuable and hitherto unseen information for consumers. Special thanks are owed to Andrew Lees and his colleagues of Friends of the Earth for inspiration, and for allowing me late-night access to their impressive library. Many others also contributed support, argument, or the seeds of inspiration, for which I am grateful. I am indebted to Sue Grant for typing the tables, to Miriam Polunin, Mari Roberts, and Katrina Whone (who subsequently joined Victor Gollancz as Managing Editor and therefore had to read the manuscript several times more) for reading the manuscript and being honest.

Thanks are due to Liz Knights, my editor, and Frances Kelly, my agent, for keeping the project and me on the rails.

Publishers are crammed with overworked and skilled staff whose contributions to the making of books all too often pass unrecognised. Other key people in Victor Gollancz who have left their mark on *Poisoned Harvest* include Ian Craig (Art Director), who designed the cover; Elizabeth Dobson (Production Director); Georgia Garrett (Senior Editor); Viv Redman (Editorial Assistant to Liz Knights); Adrienne Maguire (Publicity Director); and Levancia Clarendon (Receptionist) who remembered my name on each of the countless times I phoned or appeared in her domain.

CJR

Glossary and Abbreviations

Acetylcholine One of several different chemicals involved in the transmission of nerve signals. It allows the signal to pass from the nerve ends to the cells of muscle used in movement, to certain glands, and to the special muscle of the gut and bladder.

ACP Advisory Committee on Pesticides.

Acute effects Occur suddenly and usually pass in days rather than in months or years. Include stinging eyes, chemical burns, liver failure, etc. (See also **chronic effects**.)

ADAS Agricultural Development and Advisory Services.

ADI Acceptable Daily Intake.

Anticholinesterase Chemical that inactivates the enzyme cholinesterase (see **acetylcholine**).

Arable crop Crop grown on ploughed fields as opposed to natural meadows or forests.

ATP Adenosine triphosphate, a substance for storing and transferring large amounts of chemical energy in living organisms.

BAA British Agrochemical Association.

BASIS British Agrochemicals Standards Inspection Scheme.

Benzene ring A molecule in the form of a hexagonal ring with a carbon atom at each of the six corners. It is a common component of a vast number of organic substances.

Broad spectrum A wide range of **target pests**.

CAC Codex Alimentarius Commission.

Carcinogen Substance that can either cause a cancer to begin growth or stimulate cancer growth.

CCPR Codex Committee on Pesticide Residues.

Chronic effects Occur continuously and usually over a long period. Include damage to the nervous system, allergic reactions, etc. (See also **acute effects**.)

COSHH Control of Substances Hazardous to Health.

CWS Co-operative Wholesale Society.

DBCP Dibromochloropropane.

DDD, DDE, TDE Breakdown products of the insecticide DDT.

DEU Data Evaluation Unit.

DNA Deoxyribonucleic acid is a specialised substance concentrated in the nucleus of living cells. It forms the chromosomes that contain the genes, which allow genetic features to be copied and passed from parent to offspring in reproduction.

DOE Department of the Environment.

Dose/effect Measures the changing damage to an organ like the liver as the dose of a pesticide increases.

Dose/response Measures the proportion of a population that is affected by a pesticide as the dose increases.

Drins A nickname for the organochlorine pesticides aldrin, dieldrin, endrin.

EBDCs Ethylene bisdithiocarbamates, a group of carbamate fungicides which includes maneb, thiram, and zineb.

EC European Community.

Enzyme Special protein used in living organisms to trigger reactions without themselves being altered or used up.

EPA Environmental Protection Agency.

EPTC Eptam or S-ethyl dipropylthiocarbamate.

ETU Ethylene thiourea.

FAO Food and Agricultural Organisation of the United Nations.

FDF Food and Drink Federation.

FEPA Food and Environment Protection Act.

FIFRA Federal Insecticide, Fungicide and Rodenticide Act.

FoE Friends of the Earth.

GIFAP International Group of the National Associations of Agro-chemical Manufacturers.

Half-life The time required for the concentration of a pesticide to fall to half the initial value. For DDT in soil it is 20 years.

HCH Gamma HCH (hexachlorocyclohexane) – also known as gamma **BHC** (benzenehexachloride) – is an organochlorine pesticide commonly called lindane.

HIPE Hospital Inpatient Enquiry.

Hormones Special chemicals secreted by glands or cells in animals or plants to regulate chemical reactions or processes (e.g. oestrogen, plant auxins).

HSE Health and Safety Executive.

Immune system A combination of specialised cells and chemicals that protect animal bodies from infection by disease organisms like bacteria and viruses.

Insurance spraying Spraying against a pest regardless of whether the pest is present, to reduce the chances of a future problem.

IPM Integrated Pest Management, pest control by a combination of better cultural methods, encouraging natural predators and controls, scientific forecasting of when specific pests are likely to be present, and limited use of chemical pesticides.

Isomers Chemical compounds with the same chemical composition but differences in their structure.

LC_{50} Measure of toxicity, based on the concentration of pesticide in **ppm** or **ppb**, that will kill 50% of the test fish.

LD_{50} Measure of toxicity given as the dose in milligrams per kilogram of body weight required to kill 50% of the test animals.

MAC Maximum Admissible Concentration.

MAFF Ministry of Agriculture, Fisheries and Food.

Major exposure Exposure to relatively large doses of pesticide, usually for a short period of time and resulting from, for example, spillage or drinking.

Meiosis Cell division that produces daughter cells with half the number of chromosomes as the parent cells. This produces male and female sperm and ova for reproduction.

Minor exposure Exposure to very small doses of pesticide, for example from breathing spray drift, and from food residues.

Mitosis The process whereby cells divide, making replicates, during the growth of an organism.

MRL Maximum Residue Levels.

Mutagens Substances that cause changes to genes or chromosomes, thus altering the genetic material of cells with results like cancers, sterility and inherited defects.

NACA National Agricultural Chemical Association.

NADP Nicotinamide adenine dinucleotide phosphate, a substance used to store and transfer chemical energy produced during photo-synthesis in plants.

NERC Natural Environment Research Council.

NFU National Farmers' Union.

NOEL No observable effect level.

NPU National Poisons Unit.

OECD Organisation for European Economic Development and Co-operation.

Organic Refers to the chemistry of substances containing carbon atoms; organic pesticides are derived from natural substances (e.g.

pyrethrum); organic farming uses natural products both for fertilisers and pest control.

Percutaneous toxicity Toxicity of pesticide delivered through the skin.

ppb Parts per billion.

ppm Parts per million.

Pre-emergence herbicides Chemicals used on freshly ploughed ground to kill weeds as they germinate and before they can emerge through the soil.

PSPS Pesticide Safety Precautions Scheme.

Pyrethroids Synthesised forms of the natural pyrethrum insecticide extracted from plants.

RNA Ribonucleic acid is a specialised substance in living cells which, like **DNA**, allows exact copies of itself to be produced. It is important in making enzymes and other proteins needed by cells.

RSPB Royal Society for the Protection of Birds.

Set aside System of payments and other incentives to encourage farmers not to cultivate an agreed area of their farms.

Synergism Greatly increased activity from one insecticide when in the presence of another, usually much weaker chemical. The synergist piperonal butoxide, for example, is added to pyrethrum.

Systemics Chemicals that move within the plant or animal rather than remaining on the surface.

Target pest The specific insect, plant, animal, etc., that a pesticide is designed or applied to kill.

Teratogens Substances causing deformities to the unborn foetus.

Total herbicides Have no selective power and kill all vegetation they are sprayed on.

UKASTA UK Agricultural Supply Trade Association.

ULV Ultra low volume.

WHO World Health Organization.

Withdrawal period Time that must elapse after spraying crops before they can be harvested or fed to animals.

WPPR Working Party on Pesticide Residues.

WEEDS

We help some harmless plants to die
(That come from nature's welcome grants).
We trample, raze with hate, but why?
They have their right, as other plants,
To live, to struggle hard, defy . . .

Some weeds today might rise with pride,
Tomorrow turn to better use
For man or beast; they may provide
A food, or drug, or potent juice,
Or new synthetic goods beside.

Ah! Weeds cannot plead their place
Among the plants of gardens, fields,
As rightful crops, to grow apace;
For still we cast away their yields,
And fail to see their hidden grace.

Alexis Lawrence Romanoff, from
Reflective poems from an album of lyrics 1950–60

Introduction

Like the air we breathe and the water we drink, pesticides are taken for granted. Find a single weevil in a bag of flour, and the seller may land in court; one bright-green caterpillar tumbling from a cabbage on the chopping board produces shrieks of horror and a visit to the nearest take-away. Yet the familiar unblemished fruits and vegetables, unpopulated bags of flour, and even mould-free bread can exist only because pesticides are used at every stage of food production. Most of us are unaware, or at least unmindful, of the daily sloshing, sprinkling, and spraying of pesticides on farms and elsewhere in the environment. Such relaxed attitudes are being shaken by mounting publicity about the widespread use of pesticides and by facts which challenge the comforting thought that 'if the Government approves them, they must be safe'. The wise consumer has pesticides under close surveillance.

We probably could not live without pesticides, but we *certainly* cannot avoid them. Their use is not limited to saving our food crops from destruction or keeping life-threatening pests like malaria mosquitoes and plague-carrying rats at bay. Often they are used for purely cosmetic purposes or as a minor labour-saving device. The result is that our communities, our daily lives, are surrounded by pesticides.

Growing crops and farm animals are sprayed and dipped; stored food is protected with pesticides; our railway lines, roads, parks, and footpaths are sprayed to keep weeds down. Most homes have insect sprays, and gardeners keep an armoury of powders and liquids to keep slugs, insects, and fungi off treasured crops. Even the wallpaper, timber, paint, and carpets are impregnated with chemicals to stop moulds or insect damage. Fresh flowers from the florist and potplants from the nursery have been treated. The mains water has chemicals deliberately added to kill pathogens, and a cocktail of both accidentally and deliberately added pesticides washed from farms and factories on the way. Even the family doctor prescribes specialised pesticides in the form of antibiotics and other drugs to kill the bacteria, lice, worms, and fungi which undermine our health.

An important reason for thinking more carefully about the pesticides in our lives comes from reflecting on what they do. All niceties aside, pesticides are useful because they kill. The word *pest* comes from the Latin for 'plague' and the *cide* ending from the Latin *caedere*: 'to kill'. Whether it is harmless ants in the kitchen, unsightly whitefly on house plants, crop-destroying caterpillars, yield-reducing weeds and fungi, or bacteria that threaten human health, pesticides are designed to disrupt the chemistry of life and to kill.

There are many similarities in the biochemistry of plants, insects, and humans. There should be no surprise in discovering that a pesticide designed to kill plants like dandelions may also harm or kill us. The common garden weedkiller paraquat is extremely poisonous if swallowed and there is no known antidote. Chemicals like the insecticide malathion, or warfarin, the rat and mouse bait, target organisms much closer to humans than the dandelion. Malathion works by blocking an enzyme which controls the sending of nerve signals around an insect's body. The very *same enzyme* has the *same job* in the human nervous system. We should expect such chemicals to be harmful to us also, and perhaps in amounts similar to those used against pests.

Many of these chemicals are highly complex and unique structures whose method of action against pests is sometimes only partially understood. Many synthetic chemicals are totally new substances, invented chemicals that have never been known before. It is not surprising that their effects on other organisms in the environment are little understood, and therefore mostly unpredictable, when they are first released onto the market. Most of the new pesticides being sold are not single chemicals but mixtures of two, three, or more active ingredients, usually mixed with chemical solvents, or surfactants, to help in squirting them through a nozzle and to make them stay where they settle. Of more concern, as the number of these new pesticides increases, is the lack of knowledge about how they will react with each other in the environment, and about the unplanned effects that will result.

You don't have to drink pesticides or walk into a cloud of spray to be affected. While some break down harmlessly within hours or days of application, others are designed to stay active for months while they wait for their target organism to appear. Some pesticides work by lying on the surface but others are designed to get inside the plant or animal they are protecting. Many do not break down rapidly, or are applied at excessive doses and persist as residues in and on foods and other products. This contamination of things we eat or handle supplies us with a steady dose of substances that are odourless, colourless, and tasteless, but deadly poison.

Large numbers and large quantities of pesticides have been around for a remarkably short period – since 1950 when the petrochemicals industry took off on a sea of cheap petroleum resources. Only in the last 15 or 20 years have scientists realised, to their horror, that chemicals they thought safe at first have been ticking away like time bombs throwing up disease many years after their introduction. Most safety testing is concerned with short-term or acute illness, usually with large doses of the test substances in mind. But we have learned the hard way that substances like carcinogens show no effect at all until perhaps 30 years after they have entered the body. These chronic effects are difficult to predict and very hard to prove. Carcinogens need be present in the minutest of quantities, perhaps only a single molecule, to have their effect. Pesticides like the insecticides aldrin and dieldrin, or the anti-fungal wood preservative lindane, which is used to protect timber inside homes in the UK, can cause mutations or cancers years after exposure.

Many relatively safe pesticides can be harmful if abused or if they accumulate in the food chain. There are many examples of wildlife dying or suffering disrupted reproduction because of pesticides used to kill other organisms. Honeybees, which are vital to successful cropping, are often accidental victims of sprays aimed at harmful insects like grubs and beetles on food crops. Rachel Carson's chilling book *Silent Spring*[1] chronicled the ecological disaster of DDT, at the time considered the most successful and widely used pest-controlling substance. It was cheap, apparently safe to humans, and an effective killer of insects. However, DDT was shown to resist being broken down once it had done its job. Organochlorine insecticides like DDT accumulate in the food chain because they build up in the fatty tissues of organisms that are then eaten in quantity by higher organisms, thus moving up the food chain and threatening fish, birds, and higher animals like humans. Through such an ecological multiplier effect, dieldrin was blamed for the losses of peregrine falcons in the UK during the 1960s, and, far away in Antarctica, penguins have been found to have enough accumulated pesticide residue to cause their eggshells to be soft and break.

Around 23,500,000 kg of the active ingredients of pesticides were sold in the UK in 1988:[2] nearly 420 g (1 lb) for everyone in the population. Agriculture uses 83% of the total pesticides applied in the UK, but they are also a normal part of urban and industrial life. Most pesticides are applied out of sight so we rarely consider them: on playing fields, in factories, along railway embankments, and in parks.

But it is remarkable how we also use them in our own homes apparently oblivious to the risks involved and, often, without knowing that we are handling pesticides. Little wonder when household chemicals are advertised with photographs of people spraying their gardens without even rubber gloves to protect themselves from skin contact. This is a far cry from advertisements for farm chemicals where farmers are shown dressed in as much protective clothing as astronauts. This difference reflects new laws protecting workers, but there are no similar controls to protect the home user.

Choosing to spray the garden roses is one thing, but how many people choose to have pesticides in their new carpets or in the wood for their DIY bookshelves? Pesticides are routinely included in many household paints and wallpaper pastes, but this is not mentioned in advertisements, which entice with fancy patterns and fashionable colours, or on the labels attached to these goods. There is no legal requirement to inform shoppers when pesticides have been used in household items.

But even knowing what chemicals are used, and where, does not give enough information for wise decisions. Not only is there little publicity of the specific dangers in using pesticides, but the results of tests showing their effectiveness and their toxicity to ourselves and wildlife are also difficult to discover because the data are confidential between the manufacturer and the government.

The testing procedures that new chemicals have to pass before being approved are no guarantee of safety either to users or to the environment. Although present procedures for approvals are more rigorous than ever, there remain doubts about their ability to screen out unacceptably harmful chemicals. The more of these chemicals that are used, and the longer they are in the environment, the greater grows our experience of harmful mistakes. Many chemicals are approved, or allowed continued use, even when their safety has been questioned, on the basis of incomplete or unreliable data. Such uncertainty is rarely admitted by the authorities, who encourage consumers to believe that each approved chemical has equal amounts of indisputable scientific data behind it.

The most glaring problem for consumers trying to reach opinions or make decisions about pesticides is our lack of information. We are not stupid but, like Ophelia in *Hamlet*, are 'green . . . and unsifted in such perilous circumstance'. We live in a soup of chemicals but have insufficient information about where they are, what they are for, the risks involved, or how to use them with greatest safety. Products treated with pesticides are neither labelled nor dyed pink, so we cannot tell the difference between the risks we pay for and the safety we pray for.

In a market economy, supply and demand are allowed to dictate what is sold on the basis of three rather dubious assumptions. First, that consumers are fully informed; second, that they are able to make free and unconstrained choices when shopping; and, finally, that they use their wisdom and personal income to make the most efficient purchases possible. In general, this is a rather shaky basis for complacency about the economic or social wisdom behind what is sold and bought. Where pesticides are concerned, these assumptions are less valid than elsewhere in the marketplace.

There are no other substances found in or on the food or other goods we buy, or applied to the fields, roads, and buildings of our environment that are treated with similar secrecy and lack of public consent. There is no other product marketed which is consumed involuntarily by the entire population. With the exception of nuclear energy, there is no other product with the same potential to harm both human health and the future environment of the earth.

This book attempts to provide some of the basic information about pesticides which can improve consumer awareness and make decisions easier. It explains what pesticides are and spells out how the main types work. The risks to human health are discussed and the complex but vitally important process of evaluating and approving pesticides for use in the UK is described and assessed.

The second part of the book provides a pesticides resource kit. This contains more detailed information on pesticides, their mechanisms of action, and which ones are likely to be used on particular crops, pastures, or forests in the UK. There is a table describing the main active ingredients approved for use in the UK to help consumers identify the pesticides which appear on the labels of household pesticide products, or which crop up increasingly in the media, environmental campaigns, and daily household discussions.

The subject of pesticides is controversial. The actions of industry and government raise strong objections among consumer and environmental groups while the protests of consumers and environmentalists often anger and provoke industry and government. There are different interests on both sides and still greater differences in access to the information relevant to objective discussion. *Poisoned Harvest* also suggests that any hope of achieving a reasoned debate is made more difficult by the highly political nature of the pesticides industry.

Even as the first words of the book were committed to paper, I accepted these difficulties. I have not tried to present the *definitive* case for or against pesticides. There cannot be one. My purpose was different and, to me, more important. I have tried to combine information and arguments to influence the way all of us think about pesticides. It is clear that the way they are developed, regulated, and applied must be modified just as it is clear that the opponents of pesticides will have to accept the necessity of some form of pesticides in industrialised societies.

These changes are inevitable but unlikely to be immediate and their exact nature remains unpredictable. In the meantime, as

Joseph Joubert said, ' 'Tis better to debate a question without settling it than to settle a question without debating it.'

REFERENCES

1. Carson, R. (1962). *Silent Spring.* Boston: Houghton Mifflin Co.
2. British Agrochemicals Association. *Annual Report and Handbook 1988/89.* Peterborough: BAA.

1 What is a Pesticide?

> ★ Pesticides are as old as civilisation. The Greeks used plants like thyme to fumigate their houses.
>
> ★ Most of the chemicals now used were 'invented' after 1959 and are synthetic substances.
>
> ★ Pesticides work by interfering with the same biochemical processes which keep our bodies working healthily.
>
> ★ Most pesticides belong to one of ten chemical families. Knowing about the families tells you much about the action, uses, and environmental effects of pesticides.

Pesticides are as old as civilisation and about as varied. Although the common household fly-swat and the grazing of guinea fowl in potato crops are efficient ways of killing insects, the term 'pesticide' is usually restricted to chemicals. Sulphur was used as an insecticide in 1000 BC and both the Greeks and Chinese used arsenic. Even household fumigation has an ancient precedent: the Greeks burned thyme to rid their homes of insects. In the Middle Ages tansy was scattered on the floor, where it released its pungent, fly-repelling odour when crushed underfoot, and in 1861 Mrs Beeton advised putting 'pieces of camphor, cedar wood, Russia leather, tobacco leaves, bog-myrtle, or anything else strongly aromatic, in the drawers or boxes where furs or other things to be preserved from moths are kept, and they will never take harm'.[1]

Pesticides conjure up images of complex chemicals with daunting names. Take the herbicide alachlor. It is known to chemists as 2-chloro-2,6-diethyl-N-methoxymethylacetanilide, and has the formula $C_{14}H_{20}ClNO_2$. There are simpler chemicals like sodium chloride (NaCl), otherwise known as common salt, which does simple jobs like killing moss on paths, or copper sulphate ($CuSO_4$), which is used in fungicides sprayed on leaves. Life used to be simpler.

The early pesticides were either natural substances or simple inorganic compounds. Dung, urine, ashes, mud, and various plant extracts have all been used.[2] By the middle of the nineteenth century natural extracts were being used as insecticides. Extracts from the derris and pyrethrum plants were the first examples, and both are still used today. Derris powders contain the substance rotenone from the roots of the plant *Derris elliptica*. The properties of derris were known to the Chinese long before Europeans started extracting it. The pyrethrins are a group of six chemicals found in the flowers of *Pyrethrum (Chrysanthemum) cinerariaefolium*. The flowers were originally grown in western Asia but are now cultivated in Africa and South America. Both are among the small range of pesticides considered acceptable by the Soil Association.

The wide availability of natural pesticides should remind us of an important fact – pesticides can be both useful and safe. At issue is the safety of certain pesticides both as toxic substances and in how they are used. These points will be discussed in Chapter 4.

More sophisticated use of inorganic chemicals developed in the mid nineteenth century. Sulphur washes were the first deliberate choice of a fungicide and were effective against diseases of fruit trees. Copper arsenite was tried against beetles in potato crops in the US, and the famous mixture of copper sulphate and lime, which came to be known as Bordeaux mixture, saved the European vine industry from destruction by mildews between 1880 and 1890. Most of these early pesticides were discovered by accident. Copper sulphate and lime mixture was originally used to make the grapes growing along roadsides taste so unpalatable that they would not be pilfered. An observer noticed that vines which had been washed with the mixture were free from the dreaded grape downy mildew (*Plasmopara viticola*), which, first spotted near Bordeaux in 1878, was spreading rapidly and destroying the European vineyards. Before this dramatic chance discovery there had been no protection against the disease and the entire wine industry was under threat. Bordeaux mixture afforded another chance discovery when it was noticed that splashes *beneath* the vines killed some plants but did not harm others. This led to the development of herbicides.

It was a time of realisation and experimentation as the potential of pesticides unfolded. Scientific discoveries in the bacterial and other biological causes of human and plant diseases helped focus experiments on the killing of specific pest organisms. This was the era of Pasteur and Lister. Great efforts went into developing pesticides that would kill pests without damaging valuable crops. Enthusiasm went untempered by a sound knowledge of either chemistry or toxicology. Since common poisons like arsenic and cyanide were known to kill pests, they were used widely, even on food crops. Poisonings from zealous applications and the residues left on foods inevitably followed. By 1915 the first public concern with harm to humans from the careless choice and application of chemicals struck home when the poisoning of farm workers became obvious. The search began for safer alternatives.

This was the beginning of a major shift into organic compounds. Life is full of ambiguous words like 'natural' and 'farm fresh'. 'Organic' is another. It originally signified 'living' substances of animal or plant origin as opposed to earth materials which were inorganic, but developments in analytical chemistry then allowed scientists to copy the constituents of living things in the laboratory.

Now, organic chemistry is simply the study of compounds which contain carbon atoms, irrespective of whether they are found in living organisms or have been artificially synthesised. The consumer should note the distinction between the 'organic' in organic farming and that in organic pesticides.

The skills necessary to make synthetic organic compounds improved rapidly. New chemicals came in families. (See pages 18–22; Appendix B.) As the pesticide properties of new synthetic molecules were discovered, chemists set about 'tweaking' the edges of these molecules to make small changes with possible new and useful effects. Chemists would modify known compounds by adding different atoms like those of metals or halogens (chlorine, iodine, bromine, fluorine) to the basic organic structures of rings or chains of carbon atoms. This could be done in hundreds of ways and meant testing these totally new compounds in the hope of finding some pesticide activity. Most would have been useless, but the tiny proportion of hopefuls would be developed further as potential commercial products.

During the 1930s, fungicides like the dithiocarbamates and chloranil were developed, and they, or derivative chemicals, are still used today. By 1944, the pesticide that became both the most significant breakthrough for synthetic chemicals and the most notorious environmental poison had been patented. Chemists gave it what they term a 'trivial' name – dichlorodiphenyl-trichloroethane – but it is more widely known as DDT.

DDT was the first synthetic organochlorine insecticide but was soon followed by some equally notorious relatives: aldrin, dieldrin and endrin, collectively known as the 'drins'. Organophosphorus insecticides like parathion and malathion were developed out of the nerve-gas technology which provided the most horrific weapon of the First World War, and have the same reputation for being highly poisonous. The carbamate insecticides first appeared in 1947 and became the successors of DDT. The still widely used phenoxyacetic herbicides were developed in Britain in 1943, followed in 1958 by the bipyridylium herbicides, including diquat and paraquat. Later, families like the benzimidazoles, pyrimidines and, of course, antibiotics joined the still-growing list of new chemicals conceived by the giant international industry called agrochemicals.

How pesticides work

Living organisms are complex. Apart from the vast number and variety of chemical reactions going on all the time just for a single cell to be 'alive', the cells that make up living organisms contain a large number of structures, or organelles, that are necessary for them to tick over, grow, reproduce or move. The nucleus contains the chromosomes – the genetic code telling the cell what kind of cell it is, its shape, function, colour – in fact, everything to do with being that particular cell. There are also hollow, sausage-shaped mitochondria that store and control the release of the cell's energy stores. The Golgi apparatus is a small spherical object with the job of storing and sorting the proteins made elsewhere in the cell before sending them out to wherever they are needed in other parts of the body. Plant cells have unique organelles, the chloroplasts. These contain the green pigment chlorophyll and are arranged in layers to catch sunlight. The chloroplasts are where light energy is converted by photosynthesis to sugar which serves as an energy source for the plant's activities.

Most pesticides work by interfering with one or more of the life processes which occur within these organelles inside cells. The most common processes affected are described in more detail in Appendix A, but they can be summarised as follows.

Photosynthesis

The basic source of energy for all life on earth is the conversion of light energy from the sun into chemical energy to fuel the production of simple sugars like glucose. This is photosynthesis, and it occurs only in plants. The chemical reactions involved are complex and can be disrupted at several points. Blocking photosynthesis is the mechanism of many herbicides. Treated plants starve to death.

Energy release and transfer

Cells and organisms need energy to grow, repair damage, move, and reproduce. Plants get their energy directly from the sun, but most animals get theirs from food. Protein, starch, sugars and fats in food all provide energy. One set of chemical reactions breaks down the foods to release their energy, then specialised substances concentrate it, store it, and transfer it to where it is needed. Energy storage and transfer is analogous to trapping electricity in torch batteries or sealing propane gas in small Camping Gaz canisters, which can be stored and transported to wherever they are needed. Again the chemical reactions in cells are complex and can be disrupted by pesticides at several points. Without energy, all normal functions of cells and organisms will cease and death follows.

Nerve signals

The nervous systems of insects, fish and the higher animals and birds control muscle movement as well as the perception of environmental stimuli like light, pain, and temperature – all of which are important for survival. Nerves carry minute electrical currents along fibres which are joined at special junctions called synapses – the equivalent of plugs on household extension cables. Chemicals which interfere with the passage of the nerve current from nerve fibre to nerve fibre can either block the signal completely or turn it into incomprehensible 'static'. The result can be erratic twitching of limb muscles or the failure of breathing, which relies on muscle movement.

DNA and RNA

Chromosomes carry genes. Both are made of DNA and are part of the central computer of higher organisms. They determine that your offspring are humans and not mice or moths, and that they have the right number of eyes, teeth and other functioning parts. Chemicals which disrupt DNA can stop the reproduction of organisms or alter their performance – male insects can be made sterile, for example. RNA, or ribonucleic acid, is the second part of the computer. It is found in cells and is important for making proteins. Vital enzymes and hormones are proteins, and, since these control the rate and timing of important activities in cells, any interference with their manufacture can disrupt the cell's ability to function normally.

Growth regulators

The rate and direction of growth in most organisms are controlled by hormones. Thyroxine from our thyroid glands is necessary for achieving a normal height. Plants have growth regulators which control the rate of stem growth and also keep control over the different rates of cell growth in young leaves and stems. Too much hormone causes excessive and chaotic growth, and plants can die from exhaustion. In insects, chemicals can interfere with the hormones which control the moulting process as larval stages develop into mature insect forms.

Inventive pest control methods

Insect 'perfumes', or sex attractants, provide a clever means of luring male insects to some controlling device. These 'perfumes' are known as pheromones, and are released in the smallest of quantities but are detectable by males over 2 km away. Synthetic hormones have been developed and tried successfully. Although they are little understood, there is evidence that a wide range of organisms, including humans, make use of pheromones.

Another imaginative mechanism is sterilisation. This has been tried with some insects, using irradiation or chemicals to sterilise males. The idea relies on having male pests that still want to mate when sterilised, females who are satisfied after mating with a sterile male, and a very large number of sterilised males to release into the target population. Despite the apparent simplicity, there have been few successes and several disasters in developing practical insecticides based on this principle.

Diseases are good natural control mechanisms and have been tried as pesticides. Bacteria like *Bacillus thuringiensis* have been used for many years to infect and kill the caterpillars of the *Lepidoptera* order (butterflies and moths) of insects. Viruses have been used to infect insects, but the most outstanding 'success' was the introduction of the viral disease myxomatosis to combat the plague of rabbits that almost devastated the Australian pastoral industries in the 1950s. This was an interesting case, as the rabbits had been introduced from Britain by the early settlers, and the virus was introduced to the rabbits via the mosquito as a carrier.

Insects have also been used to control plant pests. A Mexican caterpillar, *Cactoblastis cactorum*, was introduced into Australia in the 1950s to control the spread of prickly pear over vast areas of sheep pasture. This was one of the early examples of successful biological control. The caterpillar ate only the cactus and was so voracious that the prickly pear was wiped out.

Classifying pesticides

There are too many chemicals on the market to attempt a memory job, but it is important to get used to the different types of pesticides and the various ways of classifying them. Then you will be better able to estimate the likely functions, modes of operation, health or environmental effects of any given pesticide, particularly since product labels are insufficiently informative. They normally boast the brand name – which may be something exciting but irrelevant, like Rambo, Ripcord, Spitfire, Missile, or Javelin – but do not give away the real function, much less the name of the chemicals doing the business. The ingredients will be on the label, usually in smaller print, but may include only the name of the active chemical and not the solvents or other chemicals in the product. There may be several active chemicals in the product. The table of active ingredients on pages 204–285 tells you more and, where appropriate, notes the family group to which a chemical belongs.

There are many forms of classification. For example:

- **Natural or synthetic** Natural pesticides such as derris and pyrethrum can be effective while remaining free of many of the more serious environmental hazards of some synthetic chemicals like DDT.
- **Organic or inorganic** (see pages 11–12)
- **Surface acting or systemic** Surface-acting chemicals remain on the surface of plants, soil, or other substances and objects, and go to work when the organism to be killed comes into contact with them. The systemics are absorbed into the tissues of the plant or animal being protected and spread throughout the body, particularly in plants. (Antibiotics are one example of a systemic pesticide.) The advantages of systemics are obvious: they cannot be washed off once absorbed through the outer layer of cells; they can reach areas such as the roots which surface leaf sprays cannot; they can kill organisms (e.g. fungi and bacteria) which have penetrated inside the plant; they can eradicate pests that start to grow or eat their way in, e.g. germinating fungal spores, fungi, caterpillars or sap-sucking insects. But

systemic pesticides also have the disadvantage that
they are impossible to wash off treated crops.

● The most useful means of classification for the purpose
of this book are by **function** and by **chemical family**.

Function

The majority of pesticides used today are insecticides,
fungicides, or herbicides, but there are many other uses which
should not be overlooked.

PESTICIDE	USED TO CONTROL
herbicide	plants
fungicide	fungi
insecticide	insects
nematicide	nematodes
rodenticide	rodents (mice, rats, etc.)
bactericide	bacteria
arachnicide	spiders
algicide	algae
acaricide	mites and ticks
molluscicide	molluscs
avicide	birds
slimicide	slimes
piscicide	fish
disinfectant	general bacteria, fungi
growth regulator	growth of plants
attractant	insects by attraction
repellent	flies, fleas, moths, etc.
defoliant	leaf drop
chemosterilant	insects by sterilisation
desiccant	plants by drying leaves

Knowing the function of a pesticide gives many clues about its
nature. For example, rodenticides kill rats and squirrels, which
are mammals, as are pet cats and dogs and, indeed, as are we.
Rodenticides are likely to be dangerous to *our* health.

Insecticides often work by blocking the nervous system
or energy-transfer mechanisms. It is likely that exposure to

insecticides will affect our nervous system or metabolism. All insecticides should be used very carefully around humans.

Attractants, on the other hand, have no insecticidal function but serve only to bring the target pest into contact with an insecticide. So an attractant is not likely, in normal use, to be harmful to humans.

Chemical family

Chemicals are allocated to families on the basis of their formulae, and members of one family often have similar properties. The family type gives an idea of the individual chemical's function, behaviour in the environment and toxicity to humans. Media reports and articles on pesticides tend to use chemical family or group classifications as much as the trade or chemical name for specific chemicals. Learn about a family, and you will have a good idea how either a new chemical or an unfamiliar brand of pesticide may behave.

There are many possible families to include in a classification and no universal agreement. Many chemicals can be placed in more than one family. The classification used here gives an idea of the larger families into which the most commonly used pesticides fall, and it covers the main types of chemical, where health and environment effects are of interest. The families don't account for every available pesticide because some don't belong in any of these families, but the omissions are the exception. The main families are described in summary below and in much greater detail in Appendix B. The table of active ingredients on pages 204–285 identifies the family type of each entry, unless it does not fit into the family classification used in this book. This allows you to learn more about how individual pesticides work and behave in the environment. The table also lists significant brand names of products containing pesticides mentioned throughout the book, but may not include brand names which are mentioned elsewhere but are no longer on sale.

Organochlorines
These are based on the benzene molecule with one or more chlorine atoms attached. The family's function is as a contact

insecticide and probably works by disrupting nerve signals. This explains why treated insects writhe and jerk as they die. DDT was the first developed and is one of the most persistent chemicals in the environment. Other members include lindane (HCH) and the 'drins' – aldrin, endrin and dieldrin. Lindane is used in shampoos against lice and scabies, but aldrin and dieldrin have been banned in the UK since 1989, while endrin has been banned for much longer. Because of their persistence in the environment, their residues will continue to be found in the food chain for some years to come.

Organochlorines are broad-spectrum insecticides, meaning they kill a large range of insects, which may include harmless or beneficial insects as well as the target pest. They are also harmful to fish, birds, and mammals.

The entire family is characterised by its persistence. Since these chemicals are not easily broken down in soil or water they accumulate, not just in the environment but also in the fatty tissues of insects, birds, fish, and mammals. This causes the food-chain multiplier effect, with doses becoming concentrated the higher up the chain they go. Rare birds of prey still die each year in the UK because of this effect. In humans, the accumulation may give rise to chronic poisoning, causing kidney and liver damage. Long-term build-up may also lead to symptoms of acute poisoning if fat reserves are mobilised, for example, with the sudden depletion of excess fat during dieting or a period of psychological stress.

Some organochlorines are highly toxic to humans and all are carcinogenic, which is the main reason for their being banned, after years of campaigning, in many countries.

Halogenated hydrocarbons
These are straight-chain molecules which may have chlorines or the related chemicals, bromine or fluorine, attached. They vaporise easily and are often used as fumigants. Methyl bromide and ethylene dibromide are used as insecticides, the latter being used to treat herbs and spices. The group also includes herbicides like dalapon. The herbicides damage proteins inside plants, which means enzymes are destroyed. The insecticides interfere with the nervous system. They are dangerous to humans, causing damage

to the nervous system and kidneys, and are potent mutagens and carcinogens. Because they vaporise readily and are very soluble in fat, it is easy for these chemicals to get inside our bodies, even through protective clothing. Most are now banned in the UK.

Carbamates

Concern with the persistence of the organochlorine DDT as early as 1950 led to the development of this family, which is much less persistent. It includes herbicides, insecticides, and fungicides. One large group is the sulphur-containing thiocarbamates, which include the most widely used fungicides – maneb, thiram and zineb. Strictly speaking, these are dithiocarbamates, and they are also known as the EBDCs. EBDCs were the subject of an urgent review by the UK government in 1989 after they were dropped from the US lists on grounds of carcinogenity. They remain on the approved list in the UK. (See pages 66–7.)

The insecticides disrupt nerve signals, the fungicides interfere with energy transfer, while some of the herbicides block photo-synthesis in leaves. With the exception of aldicarb (one of the most toxic pesticides available), the group breaks down readily in the environment. Chemicals in the group are water-soluble and are easily absorbed through the skin or on breathing. These pesticides are very toxic to humans, and some are known to produce carcinogens when they break down in the environment or during the cooking of food.

Heterocyclics

Some of the most widely used herbicides are in this family. It includes the triazines (atrazine and simazine) and the bi-pyridyliums (paraquat and diquat). They are known as total herbicides because they kill any vegetation they contact. The introduction of heterocyclics led to the development of 'chemical ploughing', in which herbicides rather than ploughs are used to kill weeds before planting crops. They work by disrupting photo-synthesis. While most chemicals in the group have low acute toxicity to humans, paraquat and diquat are very toxic, and there is no antidote. Atrazine and simazine, both of which can cause cancers, are often found in UK water supplies, having run off agricultural land into rivers or leached down into the water table.

The government has placed both on a special 'red list' of dangerous substances over which there should be tighter controls in water supplies.

Organophosphates

Research into nerve gases during the Second World War led to the development of organophosphates. The family is characterised by phosphate structures in chains or rings, and a great variety of other atoms may be attached. Organophosphates include insecticides which work through disruption of the nervous system, e.g. parathion, dichlorvos and dimethoate. Glyphosate is a widely used herbicide. Workers using these pesticides often experience health problems, especially from low doses over months or years. The group is very toxic to fish, earthworms and bees, but most do not persist long in the environment.

Chlorinated phenoxy

Natural plant growth-substances or hormones are mimicked by chemicals in this family. It includes herbicides like 2,4-D and 2,4,5-T, which were components of Agent Orange, the defoliant sprayed on rainforests during the Vietnam War. It worked by causing chaotic growth in the growing shoots of plants. 2,4,5-T has been the target of a trade-union campaign in the UK because of evidence that it can cause birth defects and cancers in workers' families. Chlorinated phenoxy pesticides are relatively harmless to wildlife because they break down rapidly in the soil, although they can last longer in water.

Amides and ureas

Urea, which is also used as a nitrogen fertiliser, is the basic chemical in the formulae of many of these pesticides. They include herbicides, some of which disrupt photosynthesis and some of which stop seeds germinating. The group has low toxicity but has been associated with cancer-causing residues. Its members are persistent in the environment and are dangerous to fish.

Benzonitriles

These are herbicides based, like the organochlorines, on benzene, but they can contain bromine and iodine as well as chlorine. They disrupt photosynthesis and energy transfer in plants. They are toxic to humans and produce environmental problems because they are designed to persist on sprayed plants and can wash off fields into rivers where they can kill fish.

Phenolics

Disabling the energy-transfer reactions of cells, phenols have long been used against human bacterial infestations and are natural key ingredients in herbal medicines like garlic and myrrh. The commercial formulations are used as insecticides and in wood treatment, but are extremely toxic to humans and most are banned in the UK. They are very toxic in the environment and can spread because of the ease with which they vaporise.

Synthetic pyrethrums (pyrethroids)

So effective has natural pyrethrum been through history that it was inevitable that it would be copied sooner or later. The natural form consists of six active ingredients. Synthetic forms are known as pyrethroids and include pesticides like allethrin, resmethrin, and permethrin. They are non-selective insecticides which work by disrupting nerve signals. Most are safe for humans, but because they are non-selective they kill organisms other than the target pest and are particularly dangerous to bees. They are also extremely toxic to fish.

Conclusion

It doesn't take a Nobel Prize winner in biology to realise why there is good cause to be alarmed about the flood of pesticides in the environment. There are very close biochemical similarities among all living organisms, and humans are no different when it comes to the basics. The nerve-signal disruptors, for example, are just as likely to affect us as the target insects. Like weeds, fungi, and insects, we rely on the same basic chemical mechanisms to transfer energy from food into storage and then make it available for vital life processes. Those pesticides which interfere with energy-transfer mechanisms are also likely to interfere with the way human bodies work. Both normal cell division during growth and the special form of division which produces human sperm and ova involve exactly the same processes as in other organisms. Those pesticides that disrupt normal cell division are unlikely to be harmless to us.

The future can be better, but it must be accepted that most pesticides now approved for use are, by their design, as likely to do harm to other organisms as to kill their target pests. One of the most urgent improvements is to speed up the review process of pesticides which received their approvals maybe 30 years ago (see Chapter 4) and which would not survive scrutiny under the more sophisticated evaluation procedures used today.

REFERENCES

1. Beeton, I. (1861). *The book of household management*. London: S. O. Beeton, para. 2285.
2. Stakman, E. C. and Harrar, J. G. (1957). *Principles of plant pathology*. New York: Roland Press.

2 Pesticides and the Environment

★ 450 tonnes of pesticide chemicals are used in the UK each week.

★ Pesticides are brought into every home in goods like carpets, wallpaper, clothing, cut flowers, and, of course, food.

★ British Rail operates an unpublicised night train which sprays herbicide along railway embankments, even in urban areas.

★ Local Authorities and even Water Authorities use pesticides in our environment without public warnings.

★ Pesticides wander through the environment, killing friendly organisms indiscriminately.

★ Farmers and gardeners can use much less pesticide without losing yields.

Where are pesticides used?

The UK produces and uses a large quantity of pesticides. In 1988, 23,504 tonnes of active ingredients were sold by the manufacturers. That is not the weight of packaged commercial products but the weight of the *pesticide* ingredients in the final product. The value of total sales by UK manufacturers was £1024.6 million for the same year, 9% higher than for 1987.

Agriculture is the main destination for the chemicals sold, with 83% of all pesticides ending up on farms. In 1989, UK farmers

Active ingredients sold in the UK, 1988

Product	tonnes	% total sales
Herbicides	13,376	57.0
Fungicides	5,878	25.0
Insecticides	997	4.0
Seed treatments	116	0.5
Growth regulators	2,271	9.5
Others	866	3.5
Total	**23,504**	

(Source: British Agrochemicals Association
annual report, 1988/89)

spent £440 million on pesticides, 6% of their total expenditure on inputs. Of all different types of pesticides, herbicides are the most common and once again the heaviest user is the agriculture industry. It accounts for 54% of all pesticide sales.

However, pesticides are also used widely outside agriculture in ways that can have significant effects on the environment. Non-agricultural uses may account for less than 20% of the total used, but they are no less important. The risks to user, consumer and to the environment remain and may be higher depending on the specific action required and the chemicals chosen. Both where and how pesticides are applied can be as important as specific chemical ingredients in causing unwanted effects.

Although the greatest proportion of the three main groups of pesticides goes to agriculture, significant amounts end up in household or garden use. Less than 1% of fungicides produced enter the home as pesticide, but 5.2% of all herbicides go on gardens, and 26.6% of all insecticides are sprayed in the home. Household pesticides are a good example of how little we know about the pesticides around us daily. Most home users pay too little attention to what is sprinkled or sprayed about their houses despite having to breathe sprayed air and eat off, or live on, pesticide-treated surfaces. But this is not the users' fault:

- Most pesticides which are approved for home or garden use are at the less harmful end of the range, but this does not mean they are harmless.

- Household product labels convey far less information on both content and safe application than those on products for industrial use.

- Unlike the rules facing industrial users, there are no rules or regulations governing how consumers can apply pesticides at home.

- The dangers of this are made worse by the total lack of training available for consumers in safe pesticide storage, handling, and use.

- The proximity of children, food, and pets to pesticides used in the home and garden highlights the dangers. Most pesticides used at home are sprayed in the same environment where children and pets play, eat, and sleep. Avoiding contact during or after spraying is often impossible.

- The marketing of pesticides to industrial users now places great emphasis on images and messages of safety, but the risks are played down in presenting pesticides to household users. Have you ever seen an advertisement for fly spray with the user in a gas mask and protective clothing? Have you seen an advertisement for pet flea-collars warning that pets should not be cuddled while wearing the carbaryl-, diazinon- or lindane-impregnated collars for fear of contamination?

- Home users are given no idea of how little pesticide to use for effective and safe control. With prepacked aerosol sprays, they also have no idea how much chemical is emitted in a burst of spraying. Most home users spray excessive amounts, which wastes money and greatly increases the risk of harm to members of the household. Farmers, for example, not only have guidance on optimum application rates, but also have to keep the cost of chemicals in mind and thus are less likely to overspray crops.

- Contractors treating woodworm or dry rot, and Local Authority vermin control officers who spray for wasps and fleas or leave baits for mice and rats, too often fail to fully inform householders of safety precautions which will reduce their chances of being contaminated with the pesticides used.

Outside the home, manufacturing industries and public utilities are major pesticide users, though consumers are normally unaware of the presence of pesticides in the goods or services they supply. It is difficult to provide an up-to-date list of chemicals used currently in industry because new substances are being introduced every year, and manufacturers are under no legal obligation to label their products with the pesticides used in their manufacture, much less with any possible residues in their products when sold. In the following pages, individual pesticides are mentioned as examples of what industry has used, based on a Department of the Environment (DOE) survey.[1] Some of the pesticides may no longer be used and new ones may have been introduced. Those listed illustrate the range of chemicals applied.

Non-agricultural users of pesticides

Household

- Household uses include dichlorvos **vapour strips** and **flea spray**; carbaryl and diazinon in pets' **flea collars**; pyrethrum-based **aerosol insecticides**; and lindane in **ant- and moth-sprays**.

- **Garden insecticides** contain lindane, dichlorvos, malathion, and are often mixed with fungicides like benomyl, captan, and triforine. The battery of herbicides includes 2,4-D, glyphosate, MCPA, and amitrole. Metaldehyde is in **slug baits**, while warfarin and difenacoum replace the Pied Piper in **rat and mouse control**.

- The **woodwork** and treasured **wooden furniture** may be protected from fungal or woodworm attack with chemicals like lindane and tributylin oxide, while the emulsion paints on the paintwork could contain

pentachlorophenols to keep fungi off the bright
pastels.

- The **bricks and mortar** supporting the treated
  woodwork may be washed with antifungals like
  dodecylamine salicylate.

- **Wallpapers** and their adhesives contain slimicides to
  block moulds. In 1974 there were 20 different
  chemicals in use, amounting to about 500 tonnes
  each year. The chemicals enter the production in the
  70 tonnes of water used in making 1 tonne of paper.
  A large proportion remains in the paper, and the rest
  goes into the environment. Pentachlorophenols,
  dichlorophen, and benzothiazoles are also approved
  for use.

- **Carpets** are often treated with moth-proofing
  chemicals, including common insecticides like
  pyrethrums and lindane.

- **Tap water** contains residues of a large number of
  pesticides from agricultural and industrial users (see
  page 39) as well as a smaller amount from pyrethroid
  insecticides, which are deliberately added to control
  the water hog-louse in the mains pipes.

Local authorities

Considering their responsibilities for rubbish tips and public
toilets, local parks, and gardens, and household vermin control,
Local Authorities are important users of pesticides. A large slice
of their budgets goes on recreational use, which, in Liverpool
in 1988, for instance, required the use of the following specific
pesticide functions and involved no less than 30 different
products:

- selective weed killer
- selective weed killer for annual and
  perennial grasses
- total weed killer (for paths)
- contact weed killer
- worm killer

- moss killer
- turf fungicide
- pre-emergence herbicide (for shrub borders)
- aquatic weed killer
- total weed killer
- insecticides (for nursery use)

There is no reason to believe other Authorities are significantly different. Such amenity use is how the urbanised world stays green. *Newsweek* reported, in May 1988, that in the US 300 million pounds of pesticides were poured or sprayed onto lawns, gardens, parks, and golf courses each year. Considering how people use recreational spaces in the UK, it is unfortunate that Local Authorities do not display notices warning the public when and where pesticides have been sprayed. It is assumed that the potential dangers of the chemicals are recognised and that they are applied by trained workers and according to the label instructions, but some of the public statements by Council officials do little for public confidence. Following a discussion of dogs becoming ill in local parks, a London newspaper – the *Uxbridge Gazette* – took up the issue in its columns. It discovered that atrazine, a herbicide which causes irritation to skin and eyes, was being used and told readers, 'Council spokesman Roy Mills denies there is cause to worry. The Council does not use dangerous or poisonous chemicals for spraying weeds, and the chemical used on verges has been cleared by the Ministry of Agriculture. It can be bought over the counter of any garden centre and is not harmful to animals.'[2]

British Rail

Although it may not be immediately obvious, even to the observant traveller, all railway tracks, embankments, sidings, and rollingstock yards are sprayed regularly with pesticides. Most of the 10,000 miles of track and about 20% of the embankment slopes are sprayed each year. The majority of chemicals used are herbicides, but in 1974 there were 2 tonnes of rodenticide in the total of 100 tonnes of active ingredients. Herbicides are applied manually by knapsack but also by a 'spray train' that travels at night, unseen by commuters, and does not feature in any of the BR advertisements.

BR informed the Associated Society of Locomotion Engineers and Firemen that chemicals used in the 'spray train' around Watford Junction during 1986 included Tordon 22K (picloram) and Garlon (triclopyr), Residox (atrazine), diuron and Arsenal (imazapyr) while knapsack sprayers used Atlazin (amitrole and atrazine), Destral BR (2,4-D, dalapon, diuron), Tordon 22K, and Garlon. A representative of BR described their herbicides as '. . . a new breed of low-toxin weedkiller, all of which, when correctly diluted, are less toxic than sea water.' One wonders how well the labels were read. Garlon, even when correctly diluted, is irritating to skin, harmful if swallowed, and dangerous to fish, and should not be applied either near food crops or within 200 metres of ponds, lakes or watercourses. Diuron and dalapon are irritating to eyes, skin and the respiratory system. Perhaps the BR representative meant 'polluted British coastal sea water'!

In another incident, following complaints from residents adjacent to BR lines in south London, the area civil engineer wrote, assuring them that 'The chemicals used are Tordon 22K and Garlon, which are harmless to both people and animals. I trust the contents of this letter will allay any fears which you or your neighbours might have.'

It is usually difficult to say which pesticides have been sprayed following a complaint or, indeed, on any length of track. However, BR's approved list of chemicals for national use in 1983 included:[3]

For verges	picloram, triclopyr, 2,4-D, phosamine ammonium
For track, yards, etc.	atrazine, simazine, aminotriazole, bromacil, diuron, picloram, 2,4-D, MCPA, 2,3,6 TBA, glyphosate

British Waterways Board

Herbicides are applied around canals and watercourses to remove blocking weeds. Chemicals include sodium chlorate, dichlorobenil, and diquat, but relatively small amounts are applied. These chemicals may be combined in commercial formulations rather than applied as single-ingredient herbicides.

Electricity industry

Clearing vegetation from around pylons and other installations, and keeping substations weed-free involves herbicides like sodium chlorate, iron sulphate, aminotriazole, atrazine, bromacil, 2,4-D, and 2,4,5-T.

Coal industry

Above ground, the range of herbicides is similar to that used in the electricity industry. Mines have insect problems, especially with cockroaches, and the insecticides lindane, pyrethroids, fenitrothion, and malathion are likely. Rats and mice are controlled with warfarin.

Manufacturing industry

Among the vast range of manufactured goods are many with significant amounts of perishable components, to which pesticides are applied.

- The **paper** industry has been mentioned already (page 28).

- **Natural-fibre textile** manufacturers use antifungals, of which pentachlorophenyl laurate used to be especially common; over 350 tonnes annually were used in the mid-1970s. The fungicide not only has to kill and be persistent, but must also adhere to the textile to retain efficiency. The antifungals penta- and tetrachlorophenol have been used in the manufacture of textiles from cotton, flax, and jute, and with products like fabric, cordage, and tentage. Copper and zinc naphthenates have been effective on rope and sacking.

- **Imported leather** or leather goods may be treated with chemicals like naphthenates, lindane, dieldrin, sodium arsenite, and even DDT, although this is not permitted on UK-made leathers. The difficulty is in knowing where the leather comes from, even if the finished product has been made in the UK.

- **Adhesives** for wood, paper, leather, and other surfaces are often based on cellulose, starch, and milk protein, or casein, and they are protected with fungicides and insecticides, including pentachlorophenol, tributylin oxide, sodium metabisulphite, and borax. Even the water-emulsion adhesives based on polyvinyl acetate and polyvinyl alcohols are susceptible to attack and contain pesticides.

The environmental assault

The term 'environment', like 'green' and 'road fatality', suffers from a type of sensory overload. It is used so often in the media, on soap packets, and in politicians' soapbox speeches that its meaning and importance are diluted if not lost. The following quote is a gentle mind tonic. It should be rolled around the brain, savoured and stored for reuse whenever we feel lulled into complacency about the battering dished out to the environment. On World Food Day, 1989, the Director of the UN's Food and Agriculture Organization (FAO) was moved to attack agriculture, the source of world food and guarantor of global nutrition. It is, he said:

> a victim of the ecological disaster, [and] becomes a fearsome agent of the same process as it consumes huge quantities of chemical products, particularly fertilizers and pesticides, and dumps them back into the environment. Doping has been pursued relentlessly in the sports world, and yet mankind continues its lethal self-poisoning as we systematically and freely dope the very plants and animals which are our only sources of food. To destroy insect pests, we wipe out all insects, both good and bad, causing severe disturbances along the entire chain of animal life. At the very time when the big nations of the world are finally contemplating the prospect of abolishing chemical weapons, we are waging a senseless and total chemical war on nature.

This book began with the assertion that pesticides are not only useful but essential. They are necessary for human survival and useful in maintaining acceptable lifestyles. However, all useful things can also be harmful, if used carelessly. Of course pesticide abuse is likely to cause damage, but most of the environmental dangers from pesticides come from the combined effects of *approved* chemicals being applied according to *approved* instructions (see Chapter 6).

Despite the 'wonder of technology' arguments wrapped around new pesticide products and the claims for their benefits and safety, most pesticides are blunt instruments.

Environmental damage from pesticides is dependent on two main factors: how pesticides are used; and how they behave once let loose in the environment.

Careless killers

Most pesticides are blunt weapons. Using them is like inviting a motorised road-sweeper into the dining room to collect the crumbs from under the table, or amputating a foot to remove the pain of a bunion. Both may work but at the undeniable cost of significant additional damage on the way.

Pesticides kill organisms other than the target pest. It is necessary only to read a set of data sheets for internationally available pesticides to see the potential scything of organisms. Whether for fungicides, insecticides, or herbicides, the reports of toxicity tests often record significant toxicity to one or more of the following: bees, quail, fish, mammals, dogs, mallard ducks, etc. These organisms are mentioned only because they happen to be used in the toxicity-testing programmes required for the approval and licensing of new chemicals. There is no similar record of the *total* range of species affected because such observations are rarely, if ever, made. Bats, for instance, are particularly vulnerable to lindane and dieldrin, and the use of these pesticides as a treatment for roofing timbers is thought to be responsible for the decimation of the UK bat population. Lindane is still used in these products.

Insecticides readily decimate the local insect population on sprayed crops, taking along not only the natural predators of the target pest, but also harmless, and even beneficial, insects. Honeybees get clobbered when orchards and other flowering crops are sprayed. But they are also at risk when aphids are sprayed on crops no longer in flower. Bees gather the sugary secretions from the tails of feeding aphids and can either pick up spray aimed at the aphids or get caught in the spraying process. It matters little whether toxic insecticide or herbicide is being sprayed. Strong odours from pesticides can mean bees returning to the hive being killed by guard bees who identify them as foreigners. The British Agrochemicals Association (BAA) reported a minimum of 353 bee colonies affected by pesticides during 1988 in the UK.[4]

Fish have an unfortunate skill in extracting the tiniest amounts of pesticides from water through their gills as they take their oxygen supplies. In one mishap in Canada, 70,000 trout died when herbicides were sprayed along a roadside ditch and ran off into a trout farmer's water supply.[5] Toxicity tests repeatedly show the sensitivity of fish to the complete range of pesticides. Few chemicals are sprayed directly onto water. Runoff from fields, contamination of drainage ditches when spraying weeds on the banks, seepage through soil, and irresponsible disposal of empty pesticide containers are efficient ways of contaminating water. It is estimated that 88% of modern pesticides affect fish.[6]

One fish, salmon, is treated with the pesticide Nuvan 500 or Aquagard (dichlorvos) to kill the sea lice that grow on farmed fish. While the salmon may benefit from the loss of lice, there is concern that crustaceans and other shellfish living on the floor of the lochs may be suffering. So recent is the boom in salmon farming in Scotland that there is insufficient research data to say how serious damage to other life in the lochs may be.

Plants and other wildlife around sprayed fields can be affected by crop sprays. Both herbicides and insecticides are often applied in mixtures, with several ingredients of each type of pesticide giving a wide kill-range. This has disturbed the ecology around fields, especially the large fields of East Anglia where hedgerows have been grubbed out, and the headlands dividing fields are narrow refuges from the plough-and-spray boom. Butterflies decline through the combined effects of direct attack from pesticides and the reductions of natural feed plants for the larval stages.

Many new agricultural sprays use their broad spectrum of target pests as a selling point. The herbicide Invicta (bifenox and isoproturon) is promoted as giving 'super league performance . . . It's outstanding against Black-grass and over 40 other weeds in winter cereals.' Another cereal herbicide is called Harmony M (thifensulfuron methyl and metsulfuron methyl), demonstrating how loosely the English language can be used in the service of marketing. It boasts of the 'broadest spectrum of weed control available for winter wheat. On its own it takes out over-wintered Mayweed and Chickweed, Speedwell, Panes and Forget-me-not, as well as spring germinators like Knotgrass, Redshank, and Black-bindweed . . . Quite simply it's the broadest spectrum control in its field.' But there is evidence that such total control of weeds may not always offer an advantage to the farmer. Work with rape in Scotland and on the Long Ashton Research Station in England showed that there was no loss of yield with as much as 30–35% of the ground covered with broad-leaved weeds. As the rape crop grows, it shades out the lower growing weeds.

The soil, which feeds and physically supports food crops, is not mere 'dirt' but another world rich in living organisms vital to the survival of life above ground. Soil has complex interconnections between its physical and chemical components which provide structure, the 'nutrient glue' humus and a rich population of organisms. There are bacteria in numbers approaching 3 billion per acre or, for those readers more visual than numerate, around 1.5 tonnes of live bacteria per acre. There are also actinomycetes, close relatives of fungi, and large numbers of fungi, especially in rotting leaves or other vegetation. These produce wispy white mycelia through the soil and occasionally send surface-wise their fruiting bodies known as mushrooms and toadstools. There are abundant algae, and mobile organisms called protozoa, which are single-celled animals including the amoeba. Larger and visible to the naked eye are the mites, springtails, earthworms, slugs, ants, and termites. In all, there might be around 25 million of these larger creatures in each acre of land, including 1.5 million earthworms.

The combined activities of this population include: producing nitrogen and carbon dioxide for higher plants; breaking down for reuse the vast quantities of plant and animal debris falling onto the soil each year; and keeping the soil structure suitable for air and water to move freely.[7] Most of these organisms live in the top six inches of soil. Earthworms in particular are killed by pesticides like carbofuran, which aims at soil nematodes, and the leaf fungicide benomyl. Benomyl also kills beneficial soil fungi, which, in turn, are vital food for a predator of aphids. Cereal fungicides applied to protect crop leaves have been shown to damage the soil organisms and cause additional pest problems for the crops. The fungicide washes into the soil and kills soil fungi vital to the local insect population's food chain. A key predator of the aphid is thus starved to oblivion, and the aphid population explodes, causing farmers to apply additional insecticides and fungicides.

Insurance spraying

Many pesticides are applied in far greater numbers and quantities than are necessary to achieve appropriate control of pests. A large-scale cereal farmer not only has a large potential return in 1,000 hectares of waving grain, but also a large bill for each passage over those hectares to spray weeds, insects, or fungi. There is an understandable preference for 'insurance' spraying in the hope of the greatest number of likely pests being attacked for the longest time with the smallest number of sprayings. This means more chemicals, a wider range of target pests and often higher-than-advised doses, just to be sure. Cosmetic sprays on perishable crops invite unnecessary spraying. Fruit and vegetable growers face demands for 'blemish-free' produce and risk rejection of an entire harvest if blemish tolerances are exceeded. Whether using dimethoate on lettuce or tecnazene on potatoes, growers are compelled by the market to apply 'insurance' pesticides even when there may be no evidence of the pests or only slight damage. From the hyperbole of sales reps to the regulations on pesticide use, there is little to discourage such dangerous profligacy. Hope for the future lies in the development of new approaches, especially Integrated Pest Management (IPM) (see page 168).

Climbing the food chain

Not only can pesticides affect the food chain at any point through accidental contamination but they can also spread up through the food chain in a multiplier effect, killing organisms unrelated to the targets. Birds and mammals eat poisoned insects or smaller animals that may have eaten other animals that have eaten the poisoned insects. Thus each step up the chain finds higher organisms eating large numbers of organisms below them and, as a result, accumulating increasing concentrations of pesticides in their body tissues. Birds of prey are most susceptible. During the 1970s, there were only 10 pairs of peregrine falcon in the whole of Canada between the Rockies and the Atlantic coast.[8] In the UK, similar effects have been noted on peregrine falcons and kestrels among the carnivorous birds, but also on pheasants. The Royal Society for the Protection of Birds (RSPB) has recorded bird deaths from poisonings as rising from 58 in 1982 to 103 in 1988. Incomplete figures for 1989 are already 157.[9] Not all of these deaths are due to the multiplier effect. Some are caused by pesticides used in baits laid deliberately to poison the birds, usually birds of prey. Gamekeepers breeding grouse or pheasants for shooting are often found responsible. In 1987, the toll included 2 golden eagles, 1 peregrine falcon, 12 buzzards, 6 rooks, and 17 crows.

At the top of the food chain are humans, and we are as vulnerable as other animals to off-target poisoning. There is evidence (see Chapter 7) that our health is affected by residues in food as well as by direct contact with pesticides in their manufacture or application. The general background contamination by pesticides undermines our health no less than that of wild pansies or falcons. Persistent chemicals like DDT are found in fat tissue of all humans tested, even in newborn babies. Human breast milk might be seen as the very top of the food chain. The level of DDT in American farm workers was found to be 23 ppm (parts per million) in 1972, and the average US citizen had 8 ppm. Just for a perspective on that level of contamination: were the flesh of Americans to be sold to imaginary cannibals in the EC, the meat would be declared unfit for consumption, as it contains 8–23 times the EC Maximum Residue Level (MRL) of DDT. UK levels of DDT in human fat may have fallen recently, but government figures show that in 1977, both men and women still had DDT levels in their fat that exceeded present EC MRL – an average of 2.6 ppm for men and 1.6 for women.[10] Some men had 15 times the legal limit and some women 17 times. Not even EC pigs are allowed to carry the amounts of DDT in body fat tolerated by humans in the UK.

Pesticides persist

The ideal pesticide would kill its specific target and then disappear without trace. Sadly, this rarely happens. Many chemicals are designed to last long enough to kill pests that are expected to make an appearance *after* spraying. When lindane is applied to household timbers, it is meant to last for years. In farming, many herbicides and fungicides are applied in the hope of lasting through most of a crop cycle – 5 to 12 weeks – or over the winter months before active growth begins in the spring. Many of these pesticides accumulate in the soil and kill soil flora and fauna. The more applications – especially if the pesticides are held on clay or humus in the soil – the more pesticides build up, and the more damage they cause.

Many pesticides last longer than necessary to do their job. DDT is the best-known example, persisting for approximately 10 years in the environment. Lindane, or BHC, can last 14 years, and common herbicides like paraquat and simazine last 3–5 years. There is usually a choice though, with malathion, for instance, breaking down in days, which explains its recommendation for use on household vegetables.

Nomadic pesticides

Wandering pesticides can cause havoc. Few become immobilised, like the herbicide atrazine, on contact with the soil. Many travel long distances, invisibly, appearing not only where least expected but where they may never have been applied deliberately. The longer a pesticide survives, the further it may travel.

The water route

Water often facilitates wandering. Runoff from fields, seepage into underground watercourses, accidental spillage or deliberate dumping into rivers and drains, are the main routes. Sometimes pesticides are added deliberately to water. Pyrethroids, for example, are used to kill the water hog-louse in mains pipes. The chemical can be added at levels as high as 10 micrograms per litre (μg/l) for up to 7 days continuously. This level is 100 times the EC legal limit for a pesticide in drinking water. Contamination of underground water may be more severe in lighter soils. In East Anglia the water authority has reported extensive contamination by agricultural pesticides. According to evidence given to the Agricultural Committee of the House of Commons, 'mecocrop, atrazine, simazine, dimethoate and lindane were each found *regularly* at the great majority of monitoring sites. Dimethoate, mecocrop and atrazine were also found at 10% of ground water monitoring sites'.[11] According to a 1987 FoE survey of pollutants in drinking water, at least 298 water supplies in England and Wales exceeded the EC's Maximum Admissible Concentrations (MAC) for a single pesticide (0.1 μg/l), and 76 exceeded the MAC for total pesticides (0.5 μg/l). Seventeen different pesticides were found above the MAC, but this was probably an underestimate because the data came from the water authorities, who do not routinely test for all possible pesticides. The supplies exceeding the MAC were clustered in the South East – East Anglia, the Home Counties, London, and Wiltshire. The UK government is breaching agreed EC law in failing to reduce these levels.

Problems associated with contaminated underground resources include the greater time pollutants take to 'wash out', and an apparent delayed breakdown of the pesticides. Among pesticides found in three underground water sources at levels above the MAC was the banned DDT. Additional concern comes from the lack of data on both the breakdown products of pesticides and the new products formed as pesticides react with other chemicals in the water supply. Chlorine, which is used routinely in water treatment, can react with herbicides like 2,4-D to produce chlorophenols, which are human carcinogens. The pesticide aldicarb breaks down into aldicarb sulphoxide, which is more toxic than the aldicarb.

The FoE report concluded that there were inadequate analytical methods available for testing water for pesticides and that 'the Government has no reliable information about the full extent of pesticide contamination in drinking water, let alone the concentrations present, and cannot properly assess the potential health hazards.'[12]

The atmospheric route

Pesticides also move in the atmosphere as spray drift from ground or aerial applicators. Particles can drift thousands of metres, becoming more concentrated as their water or other solvents evaporate. The new Food and Environment Protection Act (see pages 104–8), allows aerial spraying within 200 feet of houses, schools, or hospitals, but requires a minimum of three-quarters of a mile between any nature reserve and spraying. Aerial spraying is allowed to continue in a 10-mph wind.

People in the vicinity of any sprayers can easily be exposed to drift, especially if they are down wind, however slight the wind. Pesticides are often applied in what is called Ultra-Low-Volume (ULV) applicators, which dispense a very fine mist, the particles of which can be carried for miles in the atmosphere. Other forms of application include fogging and smokes, which also produce ultra-small particles that can travel long distances.

The trade route

Industrial countries have developed sophisticated means of producing foods in one area and then shipping them to where they are eventually eaten. This movement of food involves the whole world. Ingredients like spices and herbs, fruits and vegetables, and manufactured foods are shipped back and forth from continent to continent. Putting aside the gastronomic delights this allows, it is a highly efficient transport system for pesticides. Most pesticides are used on growing food crops or during storage and transport. Known politely as 'residues', these hitchhiking pesticides travel not only from British farm to dinner plate but significant amounts also appear on imported foods in the UK and any other country.

Importing foods presents a particular danger. While one country may have strict regulations on which pesticides can be used on its home-grown crops there can be little control on what happens elsewhere. Pesticides banned or severely restricted by one government can be imported regularly on food and served up to the local residents. Despite the long-standing ban on the use of DDT in the UK, the government's Working Party on Pesticide Residues had to report in 1990 that imported pork from the Republic of China regularly exceeded the EC's Maximum Residue Level (MRL).

Pesticides interact

It seems remarkable but is nevertheless true that modern pesticide testing takes little account of the ways pesticides are mixed in practice (see page 83). The mixture may occur either in a single application or through successive applications of different mixtures on one crop, or in one space like the home.

Pesticides can also react with other chemicals in the environment, producing different and dangerous by-products. The approved fungicide maneb degrades to the toxic ethylene thiourea (ETU) or ethylene thiuram disulphide, while several pesticides, including MCPA, 2,4-D, and benazolin, degrade to carcinogenic dichlorophenols. Pesticide products contain solvents, fillers, and emulsifying agents as well as the active ingredients. These substances may also be highly toxic. However, little information is offered on the combined toxicity of the active ingredients plus the 'carrier' chemicals when products are offered for approval, and less is available on how they react once in the environment. The British Geological Survey has lent its scientific credibility to the possibility of risks from these chemicals. Its report on pesticides in Britain's underground water resources considered the pesticides to be a 'serious health hazard . . . many of them break down into toxic derivatives'.[13] The Organisation for Economic Co-operation and Development (OECD) also recognised the extent of the hazard in 1986, claiming that pesticides can 'give rise, as a normal process of degradation, to metabolites and degradation products . . . which may be more hazardous than the pesticide itself.'[14] The insecticide heptachlor is converted by microorganisms in soil to a more toxic and persistent chemical than the original. Even the persistent organochlorine DDT is changed in plants to two breakdown products, DDD and DDE, or in animals to TDE and DDE. Insects that developed a resistance to DDT often did so by knocking a chlorine atom off the DDT molecule to form DDE, which was harmless to them.

Synergism is another trick performed by some chemicals. When two synergistic pesticides are used together, their combined effect is greater than the sum of the individual effects. Piperonyl butoxide is added to pyrethrum insecticides as a synergist. DDT activity is increased in resistant insect populations if a chemical called WARF Antiresistant is added. This blocks the enzymes in insects which can convert toxic DDT into safer – for insects anyway – DDE.

However, accidental synergism can occur with pesticide mixtures. Apparently 'safe' amounts of fenitrothion, an organophosphorus insecticide, appear to react with equally small amounts of DDT, causing disproportionate liver damage, followed by brain damage which mimics a well-documented medical condition called Reye's Syndrome. The danger of synergism is that both its occurrence and its effects are unpredictable. Added to our ignorance of how any *single* pesticide may behave in the environment and our ignorance about the interactions *between* pesticides, the additional complications of synergism magnify our vulnerability.

Pesticides can encourage pest resistance

Nature is rarely stupid. As toxic chemicals are poured onto organisms the organisms often change to produce resistant offspring through mutations. This can render future applications of the pesticide useless, may tempt the use of higher and higher doses in the hope of success, and sometimes creates a worse pest problem. Environmental damage can be increased by the common response of producing commercial products with a wider spectrum of target organisms.

The pesticide rarely causes the mutation; rather, it helps speed up the selection process by removing from the population the susceptible pests and leaving only those with the newly inherited resistance. These new Adonises are left to multiply, thereby spreading the new resistance through the population. Often it is a simple mutation change, creating new enzymes which can either

break down or detoxify the pesticide. But there are also im-
pressively inventive results of mutation that leave one in un-
ashamed awe of the machinery of evolution. One of the earliest
insects to develop resistance to DDT was the malaria-carrying
mosquito. Some resistant species beat the worldwide eradication
programmes simply by refusing to enter houses where DDT had
been sprayed. In another example of resistance, the young larvae
of the apple codling moth evolved the knack of spitting out the
first bit of fruit treated with stomach-poisoning insecticides.

Resistance is seen in all pest types. In 1955 there were only 50
species of insect recorded as resistant to pesticides. This increased to 185 during the next 10 years, and, by 1975, there were
305 species known to be resistant to available pesticides.[15] Given
that the range and sophistication of pesticides was increasing
over this period, resistance was winning. Weeds have shown
stubborn resistance to the s-triazine herbicides, which are the
most widely used on field crops. By 1981, there were 29 species
resistant to this group; 37, only 2 years later; and 42, by 1987. The
impact and scale of resistance is evident in Hungary where 75%
of all agricultural land was infested with a triazine-resistant
weed, *Amaranthus* (a close relative of love-lies-bleeding), as
early as 1975. Weed resistance to paraquat, diclofop-methyl, and
trifluralin herbicides is also widespread. In the UK, the four main
broad-leaved weeds in cereals are still a pest despite 20 years'
intensive use of herbicides and the apparent improvements
which manufacturers boast of. Hop growers and nursery-stock
growers are landed with herbicide-resistant annual meadow grass.

Fungi, which are simpler organisms and which reproduce
more rapidly than higher plants or insects, develop their resistance even more quickly. The more specific the action of the
pesticide, the more easily resistance develops. In one experiment
with a common mould, *Aspergillus nidulans*, nine strains of
the mould were treated with ultraviolet radiation. Five strains
mutated sufficiently to become resistant to the fungicide
benomyl. Ultraviolet light is part of the sun's radiation and
causes similar damage in humans which may result in skin cancer. In practice, commercial crops have shown the same speed
of developing resistance. The first evidence of field resistance
to benomyl was seen in the second year after it was launched,

resistance to dimethirimol was shown in its first year, and resistance to the cereal fungicide ethirimol was shown in its third year.

UK cereals offer a salutary example of the effects of resistance. In the 1960s when organomercury fungicides showed resistance, farmers switched to ethirimol chemicals until *they* showed resistance and then to the so-called mbc compounds (benomyl, carbendazim, thiophanate-methyl), which started showing resistance from 1981. Farmers responded with more broad-spectrum fungicides and more mixtures, not always with success. The Agricultural Development Advisory Service's (ADAS) annual survey of wheat crop diseases in 1988 showed that crop infection with the fungal disease eyespot had increased by 25% on crops sprayed with mbc group fungicides. Since the service

Some common crop diseases and their resistance to pesticides

Fungal disease	Crop	Resistant to
Eyespot	wheat and barley	mbc fungicides (e.g. benomyl, carbendazim, fuberidazole, thiabendazole, thiophanate-methyl)
Stem base browning and ear blight	wheat and barley	mbc fungicides
Ear blight	wheat	mbc fungicides
Powdery mildew	wheat and barley	flutriafol, imazalil, nuarimol, prochlorazpropi-conazole, triadimefon, triadimenol, triforine
Powdery mildew	barley	ethirimol
Loose smut	barley	carboxin
Leaf stripe	barley	organomercury seed treatments
Leaf spot	wheat	mbc fungicides
Leaf spot	oats	organomercury seed treatments

(Source: ADAS)

estimated that 60% of all crops were sprayed with those fungicides, alone or in mixtures, it concluded that the national cereal yield was less than would have been achieved with *no* mbc fungicides.[16] The table opposite lists some of the common cereal diseases resistant to mainstream fungicides used on UK farms.

The response to resistance has taken several routes. First, the agrochemical industry is trying to increase the rate of development of chemicals with different modes of action. Second, aware that the life of any new product is limited to the time taken for resistance to develop, the industry is trying to vary the mix of chemicals and the modes of action in their new commercial products. With no one chemical in use for more than a year or two and with deliberate variations in the modes of action, resistance cannot develop so quickly. Manufacturers also advise using different chemicals at different times during a season to try to slow down the rate of resistance development. Farmers are encouraged to follow manufacturers' advice strictly and to follow practices like choosing resistant varieties, rotating crops, and using strict crop hygiene and biological controls. Ironically, many of these practices are just what the agrochemicals were developed to reduce or replace.

Conclusions

There is little doubt that both crop and chemical scientists are now wondering whether chemical poisons have had their day and whether the time has arrived when research effort should go into more 'natural' means of pest management and control. The potential for environmental damage is clear and great. Many examples of isolated or widespread damage have been recorded, and many have achieved extensive publicity. It is even more clear, however, that environmental effects are still being played down by the pesticide industry and other too-eager defenders of these useful but accident-prone chemicals. Too little research is conducted to discover and monitor environmental effects, and when data are presented the response from industry and government is too often a defensive dismissal.

Environmental concern is introducing new pressures into pesticide development. Improvements have made possible Ultra-

Low-Volume spraying, greater specificity, and modes of action other than poisoning. Manufacturers are getting away from the crude methods of mass screening to more selective chemical design. In the former, likely chemicals were tried on a wide range of pests to see if they worked and, if so, tests were carried out to see what else was affected. Selective design uses scientists' greater knowledge of individual pests and designs more specific means of disabling them. Thus the interest in insect growth hormones, biological control and the small differences in physiology or metabolism between closely related organisms. Two examples are dimilin (diflubenzuron), which attacks the protective shell of insects, and the sulphonylurea herbicides which block amino acid – and hence protein – synthesis in plants. Even the fly-swat can be updated, as a US company has found out. The company avoids filling houses with toxic chemicals to kill pests by the simple trick of heating the house to 66°C, using butane heaters which kill all moths, cockroaches, flies, ticks and fleas therein. As long as the canary and the cat are taken outside, there is no harm to other organisms and no residues.

While the pesticide manufacturers search for safer chemicals, there are plenty of alternatives that household users of pesticides can consider. Organisations like the Henry Doubleday Research Association at Royton Gardens, near Coventry, produce many leaflets and books on organic pest controls. This doesn't mean that no pesticides are used, but rather a more logical system of pest control is followed. The basic principles can be summarised as follows:

- **Preventing garden pests.** Techniques like correct timing of digging to reduce snails and slugs, mulching to reduce weeds, choosing pest-resistant varieties, rotating crops each year, and companion planting are designed to reduce the chances of a build-up of pests.

- **Pest predators should be actively encouraged.** Healthy gardens are well populated with hedgehogs and frogs against snails, parasitic wasps against aphids and caterpillars, ladybirds and lacewings against aphids (greenfly and blackfly). Gardens should be planned with attractant plants and shelters to encourage these living pesticides.

- **Healthy plants are less attractive to pests.** Pay careful attention to plant nutrition. Too much nitrogen or too little water upsets plants and makes them more inviting to pests.

- **Plant protection** is often necessary, and usually wise, because prevention may not be enough. Netting, newspaper mulches, cloches made from clear plastic drink-containers, and slug and insect traps are effective, especially when supported by the highly selective method of picking out by hand pests like slugs and caterpillars or infected shoots.

- **Organic approved pesticides** should be seen as a last resort. Approved insecticides include derris, pyrethrum, insecticidal soap, and soft soap. Fungicides include potassium permanganate, sodium silicate, dispersible sulphur, and copper compounds. There are also traditional pesticides made from crushed garlic or rhubarb leaves against insects, and bicarbonate of soda against fungi.

Considering the difficulties in the 1990s of making the authorities act on the hazards of pesticides, it is frustrating to realise that nearly 40 years ago none other than the British Medical Association was making loud public statements about these same concerns. Their intervention was the more significant given that this was the heady time when new pesticides were appearing and making a similar impression on agricultural disease control that sulphonamides and penicillin had made on human infectious diseases only 10 to 15 years earlier. Far from welcoming pesticides as a similar panacea, doctors recognised the threat. A leader in the *British Medical Journal* on 16 May 1953 said:

Within the last decade has been a great development in the use of chemicals for plant protection . . . Some of them are exceedingly poisonous and their toxicity prompts the question whether the benefits from their use are sufficient to offset the risk of using them . . . The use of DDT to control codling moth can result in serious outbreaks of red spider mite, because the

DDT destroys the beneficial insects that normally keep the mites in check.

While the increase in food supplies and other essential crops obtained by means of the insecticides is the justification for their use in spite of their poison risk, it is essential all those concerned with their use should enforce measures to reduce those risks to a minimum.[17]

REFERENCES

1. DOE (1974). *The non-agricultural uses of pesticides in Great Britain.* Report of the Central Unit on Environmental Pollution. DOE Pollution Paper No. 3.
2. *Uxbridge Gazette.* 31 July 1986.
3. Sargent, C. (1984). In *Britain's railway vegetation.* NERC, Institute of Terrestrial Ecology.
4. BAA (1989). *Annual report and handbook.* Peterborough: British Agrochemicals Association.
5. Law Reform Commission of Canada (1989). *Pesticides in Canada: an examination of federal law and policy.* Ottawa: Law Reform Commission of Canada.
6. Rose, C. (1986). Pesticide controls: a history of perfidy. In *Green Britain or industrial wasteland?* eds E. Goldsmith and N. Hildyard. London: Polity Press.
7. Russell, E. W. (1961). *Soil conditions and plant growth.* 9th Edition, London: Longmans.
8. Law Reform Commission. Op. cit.
9. RSPB. March, 1990. Personal comm.
10. MAFF (1982). *Report of the working party on pesticide residues (1977–81).* The ninth report of the Steering Group on Food Surveillance. London: HMSO.
11. House of Commons (1987). *The effects of pesticides on human health.* Vol. 1. Report and Proceedings of the Agriculture Committee. Second Special Report, Session 1986–7. London: HMSO.
12. FoE (1988). *An investigation of pesticide pollution in drinking water in England and Wales.* A Friends of the Earth Report on the breaches of the EEC Drinking Water Directive. London: FoE.
13. British Geological Survey (1987). *The pollution threat from agricultural pesticides and industrial solvents: a comparative review in relation to British aquifers.* Hydrogeological Research, Hydrogeological Report 87/2. Swindon: NERC.
14. OECD (1986). *Water pollution by fertilisers and pesticides.* Paris: OECD.
15. Gressel, J. (1987). Strategies for prevention of herbicide resistance in weeds. In *Rational pesticide use,* eds K. J. Brent and R. K. Atkin, pp. 183–96.
16. ADAS (1989). Cereal fungicides: the full ADAS recommendations for cereals. Supplement in *Farmers Weekly,* 3 February.
17. *British Medical Journal* (1953). 16 May, 1093–4.

3 Pesticides and Human Health

★ Over 3000 pesticide poisonings are reported each year in Britain but the real figure is probably much higher.

★ A Parliamentary committee has condemned the inadequate recording of human harm from pesticides.

★ Long-term effects of pesticides are common but are also difficult to prove, so little is done about them.

★ Many common ailments like depression, asthma, rashes, and suppressed immune systems could be due to low-level pesticide poisoning.

There is no doubt that pesticides harm human health. The World Health Organization estimates that there are 3 million severe acute poisonings worldwide each year and 220,000 deaths attributable to pesticides. It is no great comfort that only 1% of these deaths occur in industrialised countries. The main difference between the deaths in less developed countries and countries like the UK lies not in safer pesticides but in greater care in using them. More careful use of a dangerous substance does not reduce its potential harm.

The dangers of pesticides are well explained by a scientist writing for users:

It is necessary to bear in mind that all [pesticides] are biocides, and that at a subcellular level most organisms are dependent on similar chemical processes. Consequently, it is inevitable that most pesticides will continue to have adverse effects in foe and friend alike. Even those based on natural products or those directed at specific 'non-human' targets cannot be assumed to be free from risk since they are still biologically reactive and may interact with totally different targets in mammals.[1]

There are two categories of harm that must be considered. First is the acute effects. These cover the immediate effects which appear soon after poisoning. They may be caused by a major exposure to pesticides – such as breathing fumes or spilling mix onto hands or face – which could hardly go unnoticed. Or they may be caused by a minor exposure – like brushing against recently sprayed crops – which can easily pass unnoticed. Such minor poisonings are not easily linked to their cause because people may have no idea of when or where they were in contact with pesticides.

The effects of major exposure may be serious enough to result in hospitalisation or death, but they are just as likely to cause less dramatic effects such as rashes. The effects of minor exposure are usually less serious and include headaches, breathing difficulties and rashes.

Longer-term or chronic effects are the second category of harm. They are usually caused by continuous exposure to very small amounts of pesticide, often over years. Chronic effects include very serious diseases, like cancer, and also such minor complaints as skin rashes.

Acute poisoning

No one seems to know how many people are affected in Britain each year by pesticides. The Agriculture Committee of the House of Commons had great difficulty getting accurate or agreed statistics when they attempted to study health and pesticides in 1986–87.[2] In Britain there is still no central organisation to collect and analyse all data on pesticide poisonings. Instead there are at least four separate organisations and as many different methods of collecting and presenting data. These are the Health and Safety Executive (HSE); the National Poisons Unit (NPU), based in

Guy's Hospital, London; the Home Accident Surveillance System operated by the Department of Trade and Industry; and the National Health Service data on hospital patients, known as HIPE (Hospital Inpatient Enquiry).

The Agriculture Committee was told by a Ministry of Agriculture (MAFF) official that there were 11 pesticide deaths in 1980, but evidence from the Medical Research Council suggested there were only 4. The reasons for such disparities become clear when the sources of data are examined. The HSE report only incidents they investigate, whereas the NPU get their data from GPs. The Home Accident data exclude any deaths that occur out of hospital. It is generally accepted, and is no surprise, that there is widespread underreporting. That means a significant number of pesticide-linked deaths occur but are either not recognised as such or simply don't get recorded. Add to these omissions the dismal fact that the organisations don't pool their data – partly because they collect different data and use different criteria[3] – and the national statistics on human harm from pesticides are inadequate, if not useless.

Illnesses, as opposed to deaths, from pesticide exposure are even less well recorded. Most minor exposures cause relatively minor acute symptoms. These include nausea, dizziness, vomiting, abdominal pain, diarrhoea, lassitude, headache, sweating, and thirst. Most people experience one or more of these symptoms often enough to dismiss them by saying 'I'm not feeling too good today.' Few people would go to the doctor with them and, if they did, few doctors would even guess at pesticide poisoning. (See pages 286–94 for a list of pesticide poisoning symptoms.) The amount of underreporting is anybody's guess because there is no known epidemiological study of such symptoms and their links to pesticides. The Agriculture Committee quoted evidence they received that in one case in which 24 farmers felt ill after using pesticides, only one illness had been reported to the HSE.[4]

Certainly, if few cases of minor acute poisoning are seen, much less recognised by general practitioners, it is no wonder that the incidence of acute poisonings among the general population is thought to be low. Again, the recorded data vary widely. In 1985, the HSE reported 103 incidents involving pesticides on farms. Most of these cases involved exposure by breathing or skin contact. About half were members of the public affected on or near farms; the remainder were farmers or farm workers. When investigated, only 40 of the total were confirmed as pesticide poisonings. This is hardly surprising given that acute symptoms may last only hours, or perhaps a few days. The HSE began a special study, in 1989, of the poisoning of farm workers using organophosphorus sheep-dips. Acute symptoms include headaches, giddiness, nausea, muscle twitching, and blurred vision. The researchers admit that by the time inspectors get to visit a suspected poisoning, the symptoms may have disappeared, and the all-important confirming evidence of raised acetylcholine enzyme levels in the blood may no longer exist as the levels may have returned to normal.[5] Such cases are recorded as 'not confirmed', which contributes to the underestimation of poisonings.

The NPU data come from accidental and deliberate poisonings by swallowing, and they seem to confirm that only obvious and serious poisonings are taken to the local doctor. A high proportion of cases involve children and rat poison, paraquat or other garden herbicides stored within too-easy reach. About 3200 poisonings were reported by doctors in 1985. The HIPE data indicate that nearly 400 people older than 10 are admitted to hospital each year with pesticide poisoning. This is fewer than the poisonings reported by doctors because only a small proportion would be sent on to hospital.

One major source of reported pesticide poisonings is the Friends of the Earth (FoE) Incidents Reports of 1985 and 1987.[6] The latter concentrated on incidents associated with aerial spraying of farm land and collated incidents reported largely through the national network of local FoE branches. It is difficult to verify such reports because of the lack of investigative facilities. As a result of the time lapse between the incidents occurring and being reported, evidence of both chemicals and symptoms may have vanished before it can be investigated. However, even such anecdotal data are important.

The report lists 149 incidents involving 63 different pesticides. Some incidents reported did not include harm to health, but of the 71 that did, only 23 were investigated by the HSE. Aerial spraying featured in a quarter of all cases where the chemical had been identified. This is because spraying aircraft are more easily spotted than tractors and are less accurate in where their spray lands. In 1986, when most incidents reported to FoE occurred, aerial spraying accounted for only 2% of the total area of farmland sprayed, but for no less than 40% of all pesticide incidents reported nationally to the HSE.

One case in the FoE report involved a group of schoolchildren walking along a public footpath in Essex. The aircraft made several passes overhead while the children waved, but then it began spraying and the children had to run for shelter by the sea wall. No action was taken. In Sharpless Primary School in Gloucestershire, aerial spraying was suspected of causing nausea, vomiting and severe headaches in 12 people and convulsions in one child. The action was a promise to give advance warning of future sprayings.

The list of cases makes glum reading. Contrary to what those with unshakeable faith in the safety of pesticides would like the public to believe, few of the cases reported involved accidents like direct contact with chemicals through improper use. Mrs Platt was hit by a cloud of herbicide while driving down the M4 motorway and suffered streaming eyes immediately and two weeks of illness. Mrs Weaving had rashes on her face, arms and legs after picking wild flowers on local verges. Two dogs died from paraquat poisoning due to eating either contaminated grass or grass seeds, while being walked in public places. The Mitchells of Willand in Devon lost their garden plants and became ill from eating their home-grown vegetables following British Rail's spraying of an adjacent railway embankment.

The range of incidents and the small number that were followed up, let alone resulted in action, are alarming. FoE covered only those cases where people thought or knew pesticides were involved. However, it was clear from their requests for anonymity that some feared reprisals if they made reports to the authorities. In rural areas where pesticide incidents are both more likely and more likely to be noted, offending landowners and farmers usually hold local influence over jobs, if not accommodation as well. This applies to the local community as well as to farm workers. In addition, the HSE had staff and resource shortages. This meant investigations were restricted to 'priority' cases that, according to an HSE document quoted by FoE, might include examples 'which appear to have a real potential for political repercussions'.[7] But the fate of reported incidents depends also on who does the investigating and what actions are customarily taken. Incidents resulting in pollution of rivers and lakes, for example, are referred to the new National Rivers Authority, which has only recently been encouraged by the government to enforce existing EC limits on pesticides in water.

In North America, where pesticide risks are taken more seriously, surveys have shown widespread poisoning of one form or another. A Canadian government-sponsored study of farm workers in the province of British Columbia found that '55% of the workers had been directly sprayed; 79% had to work in fields which had just been sprayed; more than 25% had their living quarters sprayed; and while 7 out of 10 became physically ill after a direct spraying, less than 4% of growers obtained medical help for their workers. Over 50% of the workers exposed to pesticides reported that they suffered headaches, 44% suffered from skin rashes, 35% experienced dizziness; and 36% suffered from burning eyes. Almost 70% of the workers had no proper wash-up facilities, and over 80% had no choice but to eat lunch in sprayed field areas. The study concluded that current agricultural practices in British Columbia may result in farm workers facing widespread low-level exposure to dozens of extremely toxic pesticides'.[8] Similar risks to workers have been reported in the UK, not least by one farmer-MP, John Home-Robertson, who told the Commons of the practical problems facing farm workers using pesticides: '. . . for example, having to mix a supposedly soluble chemical in precise quantities into cold water while balancing on a ladder leaning against a sprayer in the corner of a muddy field. Have they [other MPs] ever had to eat a sandwich meal in a tractor cab with their hands covered in DDT and miles from the nearest tap, let alone proper washing facilities?'[9]

Chronic poisoning

The chronic effects of pesticide poisoning are more difficult to attribute to pesticides simply because they appear long after initial exposure to the chemicals, and they can be due to very small amounts of chemical over many years. It is important to realise that few chronic effects are attributable to such obvious exposure as swallowing or skin contact. It is the very small amounts of residue, or contamination from droplets or vapour over months or years, that are of concern, amounts so small that they are rarely noticed. Often they are impossible to identify because they are odourless, tasteless, and colourless.

Exposure may come from food, ordinary household items like clothes, pot plants and pets or from household chemicals such as those in garden sprays, flea collars, rat poisons and wood preservatives. Other sources include factories which make or use pesticides, and many other places of work where pesticides happen to be present. Because the presence of chemical traces is usually unknown, it is difficult, if not impossible, to link pesticides to any subsequent symptoms. Even when a chemical may be suspected, proving the link is extremely difficult. Individuals will have been in contact with a large number of other substances during the time it takes for chronic effects to appear.

Mutagens

The most feared chronic effects are cancers, and the deformities to developing embryos in the womb. These defects can be caused by mutations which may be due to pesticides and other chemicals. Mutations caused by mutagens are permanent and inherited changes to genes, or to the chromosomes that contain the genes. The damage can be to a small segment of a chromosome or a gene, or can affect a larger length of a chromosome. Often chromosomes broken by mutagens are rejoined by protective processes in the cell but are reassembled incorrectly. Very small mutations can occur within a gene when cells are dividing and the chromosomes are making exact copies of each other. Such small chemical changes to genes may alter a single component part of a cell's protein, but that may be enough for it to be significantly different from the intended structure. If a changed protein is concerned with directing the behaviour of the dividing cells, the tissue may grow in chaos, because the genetic instructions telling the cell how to develop or when to stop growing have been severely disrupted. This is what happens in a cancer.

Few pesticides are deliberately formulated to cause chromosome abnormality, but insect chemosterilants are one example. They are often designed to interfere with the target insect's chromosomes, causing sterility. Other pesticides, however, may be transformed within the target organism, or other living things, and produce mutagenic by-products. The widely used herbicide atrazine has been shown not to be mutagenic in the pure state but to produce mutagenic products when taken up by maize plants. In essence, it is impossible to predict mutagenic behaviour of chemicals from their formulae alone. Mutagenic pesticides in common use include benomyl, captan, carbofuran, chlorfenvinphos, cyanazine, dichlorofluranid, dimethoate, omethoate, paraquat, simazine, and thiram.

Carcinogens

Cancers are chaotic cells that have lost their essential instructions. They occur in tissues that divide and reproduce in the body – blood-forming tissues, skin, intestines, and sperm – and ova-producing organs. It is thought cancers need both an initiator and a promoter of growth. A chemical may initiate an irreversible mutation which is a potential cancer but which may lie dormant for years until it is promoted into active growth. Once promoted, it becomes a cancerous cell and grows out of control. Cancerous tissue has cells that are only loosely stuck together, so it is easy for some of these dividing cells to slip away and travel to some other part of the body where they can settle and continue growing. This is called metastasis and is how cancers spread, causing secondary growths. So far it is not known why an initiator is an initiator, nor what makes a promoter promote. Sometimes the same chemical promotes and initiates, but, otherwise, different substances are required. Promoters, of whatever type, seem to require many years of insidious presence in order to act. This is why cancers may appear only many years after the first exposure to a carcinogen.

The whole process of cancer causation and growth is still shrouded in mystery, making it difficult to establish the possibility of a pesticide being carcinogenic. It is even harder to protect against them. Most carcinogens are thought to be mutagens also, but not all mutagens are carcinogens.

Many pesticides are called either potential or presumed carcinogens on the basis of laboratory studies and experience with human populations. These are often classified with the the the quaint name of 'suspected carcinogens'. It is very difficult to extrapolate findings in animal experiments to humans, not least because the same tests often produce totally different effects in different species (see pages 77–8). Since it is still unethical to conduct controlled trials with carcinogens on humans, there is no acceptable way to test the validity of using animals to identify human carcinogens. Common pesticides thought to be carcinogens include aldrin (UK approval withdrawn in 1989), benomyl, captafol, captan, 2,4-D, lindane, maneb, mancozeb, thiram, and zineb.

Teratogens

Teratogens cause deformities to the unborn foetus, producing miscarriages as well as what are termed congenital defects. While mutagens may be teratogenic, other teratogenic agents include chemicals which damage the foetus without any mutation occurring. Chemicals involved are varied and include some medicinal drugs like certain anti-cancer medications and thalidomide as well as pesticides. They seem to work by interfering with cellular metabolism in very active growth-centres like those in the rapidly growing embryo. There is evidence that embryo defects can also be caused by defective sperm which have been damaged by pesticides. This was first recorded in men working with an early organochlorine insecticide, Chlordecone. Sperm damage also manifested itself as birth defects in the wives of Vietnamese soldiers who were exposed to the massive doses of herbicides dumped on South Vietnam by the Americans. Although some foetal defects are believed to occur naturally, it is thought that chemicals in the environment may be the cause of a significant but difficult-to-estimate proportion.

The National Institute for Occupational Safety and Health (NIOSH) in the US has identified eight teratogens from existing studies, and four possible teratogens. Common pesticides under

Possible mutagenic, carcinogenic, and teratogenic hazards among pesticides

M = Mutagen, C = Carcinogen, T = Teratogen
*Indicates pesticides no longer approved for use in the UK.

Pesticide	M	C	T
aldrin*	M[1,2]	C[3]	T[3]
allethrin	M[3,4]		
amitrole	M[3]	C[3]	T[5]
atrazine	M[6]		
azinphos methyl	M[2]	C[3]	
benomyl	M[12]	C[12]	T[12]
binapicryl*	M[4]		
boric acid	M[3]		T[3]
captafol	M[4,5]	C[7]	T[5]
captan	M[4,5,18]	C[3]	T[3]
carbaryl	M[11]	C[8,9]	T[10]
carbendazium	M[13]	C[12]	
carbofuran	M[4]	C[14]	
chloramben	M[15]	C[9]	
chlordane	M[17]	C[2,9]	T[16]
chlorfenvinphos	M[4]		
chlorpicrin*	M[4]		
chlorothanonil		C[18]	
cyanazine	M[6]		
cycloheximide			T[3]
2,4-D		C[2]	T[16]
demeton-S-methyl	M[19]		
di-allate*	M[18]	C[8,18]	
diazon			T[20,21]
dicamba	M[15]		
dichlofluanid	M[4]		
dichlorvos	M[19,24]	C[18]	T[20]
dicofol		C[18]	
dieldrin*	M[1,2]	C[2,9]	T[16]
dienochlor	M[4]		
dimethoate	M[2,9]		
dinocap	M[4]		
dinoseb*			T[22]
diquat			T[16]
disulfoton	M[4,19]		
dodine	M[11]		
endrin*	M[3]	C[3]	T[3]
ethylene dichloride*	M[4]	C[9]	
ethylene oxide	M[3]	C[23]	T[3]
etridiazole	M[4]		
fenitrothion	M[4]		
ferbam	M[4]	C[8]	
ioxynil			T[24]
lindane		C[3,9]	T[3]
malathion			T[3,25]
maleic hydrazide	M[26]	C[2,26]	
maneb		C[8,9]	T[27]
MCPA			T[18]
methyl bromide	M[4,17]	C[9]	
naled	M[19]		
nicotine	M[3]	C[3]	T[3]
omethoate	M[19]		
oxine copper*	M[4]		
paraquat	M[3]		T[3]
pentachlorophenol		C[3]	T[3]
pentanochlor		C[15]	
permethrin		C[28]	
picloram		C[9]	
piperonyl butoxide		C[3]	
pirimiphos-ethyl	M[19]		
pirimiphos-methyl	M[4,19]		
prometryn	M[11]		
propachlor			T[16]
propham		C[4]	
pyrazophos	M[3]		
rotenone		C[2,3]	
simazine	M[6,26]		
sodium chlorate	M[3]		
sodium nitrate		C[3]	
2,4,5-T*	M[3,26]	C[3,9]	T[3]
tetrachlorvinphos		C[18]	
thiometon	M[4,19]		

Pesticide				Pesticide			
thiourea*		C[9]		vamidothion	M[4,19]		
thiram	M[4]	C[8]	T[10]	zineb	M[15]	C[8]	T[16]
trichlorophon	M[2,18]	C[2]	T[18]	ziram	M[4]	C[8,9]	
trifluralin		C[26]					

REFERENCES

1. Ashwood-Smith, M. J. (1981). *Mut Res.* **86**: 137.
2. Saleh, M. A. (1980). *J. Env Sci Hlth.* **B15(6)**: 907–27.
3. Sax, N. I. (1984). *Dangerous properties of industrial materials.* New York: Van Nostrand Reinhold.
4. Moriya, M. et al. (1983). *Mut Res.* **116**: 185–216.
5. Clayton, G. D. and Clayton, F. E. (1982). *Patty's Industrial Hygiene and Toxicology.* **2**.
6. Murnik, M. E. and Nash, C. L. (1977). *J Toxicol Environ. Health.* **3**: 691–7.
7. EPA (1984). *Chemical information fact sheet for captafol.* Oct, 1984. USA.
8. IARC (1976). *Monograph on the evaluation of carcinogenic risk.* **12**. Lyons: IARC.
9. Sittig, M. (1980). *Pesticides manual and toxic materials control. encyclopedia.* New Jersey: Noyes Data Corporation.
10. Robens, J. F. (1969). *Toxicol Appl Pharmacol.* **15**: 152–63.
11. Greim et al. (1977). *Arch Toxicol.* **39**: 159.
12. NBOSH (1984). *Nordic Expert Group on Occupational Exposure Limits.* Report No. 50, *Benomyl.* Sweden: National Board of Occupational Safety and Health.
13. Fahrig, R. et al. (1979). *Chem Biol Interact.* **26**: 15–20.
14. Hoberman, A. M. (1978). *Food Cosmet Toxicol.* **11**: 31–43.
15. Worthing, C. R. (1987). *The pesticides manual.* Eighth Ed. London: British Crop Protection Council.
16. Fletcher, A. C. (1985). *Reproductive Hazards at Work.* Manchester: Equal Opportunities Commission.
17. Simmon et al. (1977). *Dev Toxicol Envir Sci.* **2**: 249.
18. IARC (1983). *Monograph on the evaluation of carcinogenic risk.* **30**. Lyons.
19. Hanna, P. J. and Dyer, K. F. (1975). *Mut Res.* **28**: 405.
20. Kimborough, R. D. and Gaines, T. B. (1968). *Arch Environ Hlth.* **16**: 805–8.
21. Hopper et al. (1979). *Science* **205**: 591.
22. Gibson, J. E. (1973). *Food Cosm Toxicol.* **11**: 31–43.
23. NIOSH (1977). *Occupational Diseases: a guide to their recognition.* Cincinnati: US Dept Health and Human Sciences.
24. IRPTC (1983). *Bulletin.* **6**: (1).
25. Sylianco, C. Y. L. (1978). *Mut Res.* **53**: 271–2.
26. Report on the Genetox Program (1983). In *Mut Res.* **123**: 183–279.
27. Chernoff et al. (1979). *J Toxicol Env Hlth.* **5**: 821.
28. USFDA (1985). *Surveillance Index.*

suspicion include aldrin (UK approval withdrawn 1989), benomyl, captan, captafol, 2,4,5-T, dichlorvos, and diazinon.

The table on pages 58–9 lists some of the chronic hazards that are established or strongly suspected from studies throughout the world. So difficult is it to establish firm proof that it is easy to oppose any listing of carcinogenic or teratogenic chemicals simply with the objection that firm, easily reproducible and scientifically 'proven' evidence has not been presented in most cases. Where there are published papers and reviews of available evidence, up to 61 chemicals are implicated as possible carcinogens. As is discussed in the next chapter, the failure of a chemical to appear on a 'danger list' does not necessarily mean it is safe. In 1978, NIOSH published a list of possible mutagenic or carcinogenic pesticides to which workers were exposed during pesticide manufacture, but were unable to assess 1300 pesticides because there were no data available.[10] Given these difficulties and the health and hereditary risks, it is more than prudent to respect lists of 'potential' or 'possible' hazards until more conclusive evidence emerges.

Other chronic effects

There are other chronic effects that are less frightening but still unpleasant and unacceptable. Included are slight, but accumulating, damage to kidneys, liver, lungs, and heart, as well as a range of nervous disorders and allergies. Allergies and sensitivities to pesticides are the result of the body's immune system recognising some unwanted chemical or foreign body and going into a protective overdrive. The chemical, or body in the case of pollen in hayfever, need not be harmful to health, but this does little to reduce the distress caused.

Pesticides may produce reactions like asthma and dermatitis of various types, and they may induce hypersensitivities that cause subsequent reactions to very small concentrations of the initiating chemicals. Once susceptible people have reacted to an acute contact with a pesticide, subsequent attacks can be brought on by the merest trace of the same chemical.

With pesticide residues a normal component of daily foods, many hypersensitive people experience symptoms like abdominal pain, migraines, and fits which could be precipitated by these residues. The first dose of pesticide could have been years before and may not have been noticed by the sufferer. The difficulty of identifying the cause is made worse by the widespread occurrence of pesticides and other chemicals in the environment and the difficulty in finding uncontaminated samples with which to test these sufferers. Of course, not knowing what residues may be on food also makes it difficult for hypersensitive people to avoid symptoms. Even organic food can have pesticide residues from background environmental contamination.

Lindane, which is used in wood treatments and also head lice shampoos, is believed to result in reactions which include epileptic fits, headaches, and also a serious blood disorder called aplastic anaemia, which may not appear until months after exposure.

The immune system may also be damaged as well as overstimulated by pesticides, as in allergies. The House of Commons Agriculture Committee received evidence from the US that the widely used soil nematicide and insecticide, aldicarb, can suppress the immune system, causing AIDS-like symptoms. The existence of AIDS is now well known, but little is known about how the immune system can be so severely damaged and of the extent to which this immune system suppression occurs in the population. It was noted by the Committee, however, that aldicarb has been banned in the state of Wisconsin because of this evidence.[11] Parathion, an organophosphorus insecticide, has been shown to suppress the immune response in hamsters after a single dose. The effect is a general lowering of resistance to common infections and could be occurring in human populations.

Delayed damage to the nervous system, so-called neurotoxicity, can follow exposure to pesticides. One of the main characteristics of long-term damage is destruction of the insulating myelin coat that surrounds nerve fibres. Organophosphorus pesticides that work by disrupting the sending of nerve signals seem to cause this long-term damage to the myelin coating. The results are muscle weakness and paralysis. There is also evidence of chronic disruption of the acetylcholine mechanism for carrying nerve signals from nerve fibre to nerve fibre, from the brain to distant action centres like the fingers or toes. Chronic effects are seen as behavioural changes, depression, and slower response times. The carbamate insecticides cause similar chronic damage to the nervous system. There are also suggestions that such diseases as Parkinson's might be linked directly or indirectly to the effects of pesticides.

REFERENCES

1. Wilkinson, C. F. (1987). Environmental toxicology: its role in crop protection. In *Rational pesticide use* eds R. K. Atkin and K. J. Brent. Cambridge; Cambridge University Press.
2. House of Commons (1987). *The effects of pesticides on human health*, Vol. 1. Report and Proceedings of the Agriculture Committee. Second Special Report, Session 1986–7. London: HMSO.
3. House of Commons. Op. cit.
4. House of Commons. Op. cit.
5. *Farmers Weekly*, 24 November 1989, p. 17.
6. FoE (1987). *Chemical trespass – whose turn next?* Friends of the Earth Second Incident Report. London: FoE.
7. FoE. Op. cit., p. 5.
8. Matsqui-Abbotsford Community Services (MACS) (1982). *Agricultural pesticides and health survey results.* Abbotsford, British Columbia: MACS.
9. Quoted in Rose, C. (1986). Pesticides: an industry out of control. In *Green Britain or industrial wasteland?* eds E. Goldsmith and N. Hildyard, London: Polity Press.
10. NIOSH (1978). *Criteria for a recommended standard: occupational exposure during manufacture and formulation of pesticides.* Washington, DC: Dept of Health, Education, and Welfare.
11. House of Commons. Op. cit.

4 Safety Testing and Approval

★ Pesticide approvals assume the chemicals are 'safe until proven harmful' rather than the wiser 'harmful until proven safe'.

★ Safety tests are incapable of proving that pesticides are harmless.

★ Many pesticides used in the UK have never gone through the compulsory approval process which began in 1986.

★ Pesticide manufacturers, not an independent body, conduct the safety tests which the government relies on to grant approval for new chemicals.

★ More public accountability is essential in the testing and approval of pesticides.

Since the 1950s, when pesticide use began to explode across the globe, there has been pressure for regulation at every step. The Environmental Protection Agency (EPA) in the US estimates there are 45–50,000 pesticides based on 600 active ingredients. The UK today has over 4000 different pesticide products on the market, using about 400 active ingredients. (See table on pages 204–85.) Since most are newly created chemicals and will be spread throughout the environment, it is essential that they be tightly controlled.

Even before 1950 there was pressure for regulation, but that was to ensure buyers were not being offered worthless or diluted products. Fraud was likely when users could not tell the difference between different products, each of which smelt 'chemical', and was a white powder. Concern with the environmental effects began in 1948 when the persistence of organochlorines like DDT emerged. This led to the formation of the International Union for Conservation of Nature linking national conservation bodies. Then came Rachel Carson's book, *Silent Spring*, which clearly presented the scale of the dangers and created the climate for mass understanding and response. International and national organisations multiplied, but the central concern was with the environmental rather than human health effects.

Slowly during the 1970s and more rapidly during the '80s, fears for health overtook environmental worries. Even though the very persistent pesticides with high mammalian toxicity like DDT were banned in most countries, public concern grew as never before, but the emphasis was on health rather than on the environment. It was only with the banning of DDT in the US in 1972 that the Federal Insecticide, Fungicide and Rodenticide Act (FIFRA) demanded more rigorous testing for the safety of pesticides instead of only their effectiveness. New pesticides had to have lower health risks as well as be good at poisoning target pests. In 1983, the FIFRA provisions began to reflect a new emphasis at the Environmental Protection Agency to concentrate on the human health aspects of pesticides, 'assuming that if these toxins were safe for us, they will then be safe for the rest of our environment'.

Before the 1980s, interest in health and pesticides emerged slowly. Since then, better observation and publicity of the effects of long-term exposure to chemicals has dramatically increased public concern. Acute, or immediate, toxicity of pesticides may be reduced by approving only those 'safer' chemicals, i.e. those which require high doses to cause immediate harm. But with chronic effects, there is no such dose relationship. Very small amounts of pesticide absorbed continuously over a longer period can cause serious and unannounced health problems. Today, despite the obvious potential for acute poisoning from almost every pesticide marketed, it is the fear of chronic, long-term effects which is dominating the public interest, because pesticides are all about us in the environment and offer sufficient and continuous contact.

The general lack of necessary information about the behaviour of these chemicals once applied is alarming. There are no reliable data on where in the environment these low doses are coming from. Food residues are almost ubiquitous, and even water supplies are regularly contaminated, but more precise information is scarce. The monitoring of pesticide residues by government is at a low level, and most of the approved pesticides are not included in routine residue analyses (see Chapter 5). Even the type and level of pesticide contamination in drinking water is not known because water is not systematically monitored for all possible pesticides. The 1988 FoE[1] report on pesticide contamination of water quoted the assistant director of Water Quality in the Environment Department of the Water Research Centre's evidence to a Parliamentary committee as saying, 'We do not have comprehensive information on pesticides in drinking water as such, and that lack of information is worrying.'

The regulation of pesticides is concerned with deciding which chemicals can be sold, what they can be used for, by whom, and in what manner. The most important element in all regulation activity is the safety of chemicals – safety for the user, for other consumers, and for the environment generally. This is also the most controversial part of regulation. The majority of pesticides are made by some of the largest corporations in the world, and some of the most important national industries, like agriculture, have come to rely on their use. In the UK, these corporations conduct all the tests on pesticide safety, and farmers have not been known for demanding closer scrutiny of their most efficient crop-protection chemicals.

With the increasing sophistication of pesticide chemistry and a growing list of potential uses, safety testing is complex. It is also secretive, with the government and the manufacturers keeping all the test data to themselves. Consumers have trouble learning of the health risks from particular chemicals. But whatever information they manage to glean on the risks, they have almost no chance of being able to decide for themselves whether a chemical is safe *enough* to use, or not. They do not have access to the official test data, and, most importantly, they are not consulted.

Pesticides are big business, and the safety testing of pesticides should be everybody's business. This chapter explains the controversies in safety testing, explains the tests that are required before a new chemical can be approved, and outlines their inadequacies. The entire approvals process is shown to be so riddled with problems and biases that it should be the subject of an urgent national review.

Safety is a matter of opinion and not fact

Safety is a relative concept. Not only is there no agreed definition of a safe chemical, but there is no simple scientific test which gives a clear-cut, unambiguous measure of 'safety'. There are plenty of data from multitudes of tests, but all judgements on pesticide safety still depend on opinions rather than scientific facts.

Before delving further into this contentious and tendentious world of testing, it is important to get three points clear. First, pesticides are used only because they are either poisons or in some other way interfere with life processes. Second, there is no scientific justification for any assurance that pesticides are harmless. In the words of one pesticide scientist, 'Honest toxicologists can never, ever provide an assurance of safety – only some low level of risk for even the most innocuous chemical.'[2] Third, the important terms, 'safe' and 'harmless', are often used in confusing and misleading ways when explaining the risks from pesticides to consumers. The everyday meaning for both terms is that *no harm, however small*, will follow an action. But when pesticides are described as 'safe' by the industry or by government officials, a very different and deceptive meaning is usually intended.

Consider what happened in 1989, when the British government asked its Advisory Committee on Pesticides (ACP) to review the safety of the most widely used crop fungicides in the UK, the EBDCs – maneb, zineb, and mancozeb. Government agencies in the US had concluded they were carcinogenic, and there was media and public debate on whether these fungicides should continue to be allowed on UK crops. When junior Agriculture Minister, David Maclean, announced the ACP's findings to the public by issuing a press release, he said that they had 'reached a clear conclusion there was no evidence of risk to consumers from the use of EBDC fungicides'.[3] This almost suggests consumers could eat EBDCs for breakfast. In fact, the Committee had *insufficient evidence* to decide whether or not there was any risk. But Maclean didn't explain that vital limitation. The misleading use of the terms is made more obvious a few lines later when he said, 'The committee has identified some instances of residues [of EBDCs] in spinach and lettuce which although *at safe levels* need further investigation' (my italics). But, surely, if they are at 'safe levels', why the need for further testing? A few lines further on, he said of those same residues that 'action should be taken to ensure that residues are kept to a minimum. This is in line with existing policy that, *however safe*, residues should be kept to a minimum by good farming practice' (my italics). So pesticides can offer 'no risk' or be 'safe', and yet their residues still need to be reduced to a minimum – surely this is because they are still harmful.

The EBDC example illustrates the most important aspect of safety testing from the consumer point of view. Even the minister responsible relies on opinions rather than facts, and in assuring consumers that pesticides are safe, he goes through verbal gymnastics to disguise the basic fact that all pesticides are fundamentally harmful.

Much of the controversy over pesticide safety hovers around either the factual basis of an opinion or whose opinion counts most. And as the EBDC example shows, the interpretation of available facts depends on an individual's general approach to safety.

Safety testing philosophy

The philosophical approach to safety is crucial to both the choice of safety tests and the response to the test results. The issue is simply stated – should pesticides be judged safe until proven harmful or harmful until proven safe?

In the US, it is up to the manufacturer, user, and others to satisfy the requirement that 'when used in accordance with widespread and commonly accepted practice [a pesticide] will not cause unreasonably adverse effects in the environment'.[4] Pesticides are presumed unsafe and manufacturers are required to show that their chemicals are safe before they are allowed to be used. In Canada, a similar approach is evident. The Minister of Agriculture may refuse to register a chemical if he or she is of the 'opinion . . . [that it] . . . would lead to an unacceptable risk of harm to . . . public health, plants, animals or the environment'.[5] Again, the onus of proof lies with the manufacturer. Both Canada and the US also have long-standing laws regulating pesticides: Canada has the Pest Control Products Act of 1939, and the US, the Federal Insecticide, Fungicide and Rodenticide Act (FIFRA) of 1947. The result has been a greater number of chemicals banned in these countries than in the UK.

British governments have followed a different philosophy, if only by implication. Pesticides are presumed safe unless shown to be harmful. Until the Food and Environment Act, 1985 (see Chapter 6), pesticide approvals were handled under a voluntary scheme, the Pesticides Safety Precautions Scheme (PSPS). The pesticide manufacturers agreed to submit data on formulations and safety tests (carried out by themselves) to the Ministry of Agriculture (MAFF), which then considered the data through expert committees and advised the companies of any restrictions, label requirements, etc., to be followed in marketing products. The striking feature of the PSPS was that it was based wholly on *voluntary* agreements between the government and industry associations. There was no attempt to verify safety tests before chemicals were released onto the market and no formal requirement to instruct pesticide users on necessary precautions.

Even since the 1947 Agricultural Act, which laid the foundations for a heavily subsidised UK farming industry, MAFF has had a history of closer collaboration with the food-producing industries than with consumers. Policy has also focused more on the farmers' good than on the national good in a wider sense.[6] Close relationships between farmers' organisations and the MAFF have been paralleled by similarly close links between the MAFF and both pesticide-manufacturer organisations and industry representatives. One review of the UK's voluntary pesticide scheme claimed that 'the approach in Britain has good working relations between government and the industry as a *sine qua non*.

The voluntary scheme came into existence in 1957 following such consultations.'[7] This relationship has favoured the 'safe until proven harmful' approach which still exists under the latest UK legislation, the Control of Pesticides Regulations, 1986 (see Chapter 6), and has delayed the introduction of tighter controls on the sale and use of pesticides.

The following example illustrates how favouring the industry can work against improvements to safety. In response to a series of deaths and illnesses among agricultural workers using organophosphate pesticides in the early 1950s, the MAFF convened an expert committee, the Zuckerman Committee, to propose solutions. Their recommendation for statutory provisions to regulate pesticide use in the UK, including the compulsory wearing of protective clothing by operators, was thwarted by civil servants in the MAFF. In an internal MAFF memo, quoted by veteran environmental and pesticides campaigner Chris Rose, civil servants claimed 'It may be the case that in this country trade organisations [i.e. the industry] have a highly developed sense of social responsibility, which enables us to achieve satisfactory control by voluntary means. The Industrial Pest Control Association would be opposed to the licensing of rodenticides and insecticides but would support a voluntary approval scheme.' The civil servants' own lack of enthusiasm for statutory controls was further evident in their reaction to Local Authority requests for a scheme to register pest-control firms and operators. Their memo went on: 'We have resisted such representation on the grounds (a) that it would be difficult to ensure the registration was complete . . . (b) that a new offence of "failure to register" would be created, (c) that the work involved would not be justified, and (d) that to impose the kind of conditions that the Local Authorities desired would be an unwarranted interference with the freedom of commercial concerns.'[8] The civil servants put interest in their own convenience and that of the industry before the workers' safety.

Because neither the full set of submitted raw data nor the proceedings of the Advisory Committee on Pesticides, which recommends approval, are fully available for study, it is very difficult to observe how the relationship with industry works in practice. However, it is possible to examine the tests required of

industry to see what information on safety is available to the
decision makers, and how it might be used to assess the safety of
pesticides.

Toxicity tests

Under the Pesticide Regulations, 1986, approvals for all agricultural and horticultural pesticides, including chemicals for
domestic use, are handled by the MAFF. All pesticides for the
preservation of wood, bricks, and ship hulls, and public health
chemicals like rat poisons are handled by the Health and Safety
Executive. Both use the same general criteria for approvals and
draw on the same set of safety tests.

Applications for approval must be accompanied by a large
amount of chemical data on the active ingredients and commercial formulations or products, plus the results of specific toxicity
tests which are designed to indicate both human and environmental toxicity. The tests might include acute and chronic
toxicity in feeding trials on animals, fish, and birds; tests of
mutagenicity and carcinogenicity on animals; tests of poisoning
by pesticide entry through the skin or lungs; tests of skin and eye
irritation on small mammals; and tests of toxicity to earthworms,
bees, fish, and other organisms in the environment.

Most of the toxicity tests are carried out on the active ingredients used in commercial products. When approval is sought
for pesticide products, which might contain two or more active
ingredients, the safety tests submitted usually cover only the
individual active ingredients. Although some tests may be required of the entire mixture, it is unusual to have a *complete*
set of toxicity tests on all the possible interactions. In addition,
more than one product may be used at the same time, or on
the same site, producing further, untested, interactions (see
page 83).

Human toxicity is decided by extrapolation from these standard animal tests, supplemented by direct observation of human
poisoning where possible, using records of health effects on
workers and on the general population. Volunteers are sometimes recruited for tests. They are given measured doses of

pesticides, and the biochemical effects are observed closely through analysing for enzymes and other constituents of the blood.

Epidemiological data on the effects of pesticides in the population are also invited as a part of the information package that may be submitted. These data might include measurements of the pesticides in the blood and urine of random samples of people, or of health among people living closest to places where particular pesticides are made, stored or used.

Not all of these tests are necessary for every new chemical. The guidance for companies under the PSPS, which is still given under the 1986 legislation, says the complete range of tests 'should not be regarded as a rigid plan to be universally applied, but only as a guide to be modified for each particular substance. It is strongly recommended that at an early stage manufacturers should discuss their programme of investigation with the officials concerned. It is not expected that all the necessary toxicological data for commercial clearance will be available during the early development of a product.'[9] The range of tests required, and any particular tests that should be emphasised, are agreed between the MAFF and the company. Discussions between company employees and civil servants begin early, and are actively encouraged by the MAFF. Detailed suggestions of how tests should be conducted are given to companies, but there is no real standardisation of test design or even, in most cases, of the specific test animals a company must choose. Despite all this 'guidance', the responsibility for providing sufficiently convincing data lies with the companies.

The tests are designed to detect two forms of toxicity – acute and chronic. (See page 50 for definitions.)

Acute toxicity tests

The suggested range of suitable animals for the tests is small. Guinea pigs and rabbits are recommended for skin irritation tests, rats for the percutaneous toxicity tests, rabbits for eye irritation, and guinea pigs for allergic sensitivity. The Japanese quail is recommended for bird toxicity tests, while the domestic hen is used in neurotoxicity testing of organophosphorus compounds or similar anticholinesterase pesticides. The harlequin fish provides the test model in fish toxicity tests, but the trout is suggested for studies on accumulation of pesticides in tissue.

These choices have been made not because the particular animal, bird, or fish enables direct and accurate extrapolation to humans – far from it. The selection is more concerned with the ease of detecting sensitivity to the pesticides in question or with the reaction being tested. For example, the domestic hen is particularly sensitive to organophosphates. Convenience also counts. The harlequin fish is chosen because it is small enough to fit 10 fish into easy-to-handle, 500-ml conical flasks. The rat is chosen for mutagen tests because it has a short life cycle and is small enough to keep the large numbers required for mutagenicity tests in the laboratory.

The most important measure of acute toxicity is the tests designed to estimate the LD_{50}. This is the dose required to kill 50% of the test animals. The pesticide is usually given by mouth (oral LD_{50}) but may be given through the skin (dermal LD_{50}) or even by injection. Reported LD_{50} values should always state the way the pesticide was administered and be expressed as the weight of the chemical in milligrams per kilogram of the test animal's body weight. The acute oral LD_{50} for paraquat with male rats is 13 mg/kg. For fish, a similar measure is the LC_{50}, or the concentration of pesticide that will kill 50% of fish. It is measured as parts per million (ppm) or billion (ppb) of pesticide. Translat-

	Oral toxicity (LD_{50}) mg/kg		Dermal toxicity (LD_{50}) mg/kg	
	Solids	Liquids	Solids	Liquids
Extremely hazardous	≤5	≤20	≤10	≤40
Highly hazardous	5–50	20–200	10–100	40–400
Moderately hazardous	50–500	200–2000	100–1000	400–4000
Slightly hazardous: unlikely to present acute hazard in normal use	>500	>2000	>1000	>4000

(Source: WHO Recommended Classification of Pesticides by Hazard: 1986–7. VBC/86.1. Rev. 1.)

ing LD_{50} values to human toxicity is done in different ways by different organisations. The World Health Organization system (WHO) is a common standard. It classifies pesticides in one of four toxicity categories as shown in the table on page 72.

A more useful classification which relates toxicity to an easily visualised human dose is represented by the following table:

	LD_{50}	Amount to kill a 150 lb/70 kg man
Super toxic	<5 mg/kg	<7 drops
Extremely toxic	5–50 mg/kg	7–100 drops
Very toxic	50–500 mg/kg	1 tsp–1 oz
Moderately toxic	0.5–5 g/kg	1 oz–1 pt or 1 lb
Slightly toxic	5–15 g/kg	1 pt–1 qt
Practically non-toxic	>15 g/kg	>1 qt or 2.2 lb

(Source: Grosselin, R. E., Smith, R. P., Hodge, H. C. (1984). *Clinical toxicology of commercial products.* Baltimore and London: Williams and Wilkins.)

The acute data, especially the LD_{50} values, are often the only toxicity information presented when safety data on pesticides are published, and should be treated with caution. They may well be accurate data based on the test animals used in controlled laboratories, but they reveal little else. They show toxicity, but do not say how death occurred, whether quickly or over several painful hours, and they say nothing about the state of health among the remainder of the population. In fact, by their definition, they give information only about deaths that occur rapidly. They give no clue at all to the effects on health of lower doses over longer periods of time. An LD_{50} in the WHO 'slightly hazardous' range may be relatively harmless in acute poisoning, but could produce serious long-term health damage.

Immediate and short-term effects
Among tests for acute toxicity, some tests look for immediate effects while others are concerned with the toxic effects which might occur only after several days.

Short-term toxicity tests look for damage to organisms that does not kill immediately, as acute poisons do, but may cause a slowly developing illness and eventual death. Short-term deaths are expected within a matter of days or hours. With fish, for example, deaths may be recorded in each of 24, 48, 72, and 96 hourly periods. The most common test animals are usually rats, which are kept for up to 90 days. Regular samples of blood are collected to test damage to organs like the liver, and animals are killed regularly during the test period to study any changes to their organs at autopsy. Different doses of each pesticide are given to separate groups of rats over the test period. Short-term deaths are often caused by damage to organs, like the liver, which may no longer be able to perform their essential tasks. The liver is often affected by poisons because one of its jobs is the accumulation of toxins for conversion into safer chemicals that can be passed out of the body in the urine.

One of the tasks of the toxicologist is to identify any dose/effect relationships which apply to humans. This is the degree of damage to an organ like the liver as the dose increases. Another form of dose relationship is the dose/response, which is concerned with the proportion of the population showing a particular sign of damage as the dose increases. It is accepted that, in the case of pesticides, there will be some level of contamination of either the workers or the general environment, but it is useful to identify the minimum dose which produces an effect. This is defined as the NOEL (no observable effect level) and is used to decide the 'safe' level of environmental contamination or food residues (see pages 119–20). It is often referred to as the 'no effect level', conveniently omitting the all-important 'observable' and implying, possibly incorrectly, that the pesticide is harmless. However, because an effect is not observed, it cannot be said not to exist. Its occurrence may not have been observed during the period concerned or it could be a relatively infrequent effect and did not occur in the small number of animals studied.

Chronic toxicity tests

Chronic or long-term tests are aimed at finding mutagenic, carcinogenic or teratogenic pesticides the effects of which are not normally evident until several years after exposure.

Tests for carcinogens normally use the rat or mouse. With the rat, tests are run for 2 years (18 months for mice, which have shorter lives) to see if such diseases appear during a whole life span. Typically, groups of 50 animals are used for each test dose. With these tests toxicologists are, in fact, able to look only for serious or highly active carcinogens. The rats are given doses which range from a high dose that should produce some toxic effects to a low dose that is estimated to be equivalent to the maximum amount of a pesticide that a human is likely to consume. The high dose is also chosen to help identify high-risk carcinogens in the small groups of animals used but may miss identifying much lower-risk carcinogens, which may produce fewer cancers.

Although one person in three can expect to develop a form of cancer, certain forms of cancer are relatively common while others are rare. In 1974, only 19 in 1000 British men would develop bladder cancer during their lifetime. This means that in a random group of 100 men, only 2 would be expected to develop bladder cancer. Many cancers caused by chemicals in the environment are even less common and may be expected only once in 10,000 people. Using rats, and assuming the tests are an approximate model of how cancers would develop in humans, there would need to be at least 10,000 rats in a test before 1 cancer would be expected to show up during the 2-year life of the rats. In a laboratory test with 50 rats, the odds against finding that one cancer are 200 to 1.

Testing for mutagens is also important because while all carcinogens are also mutagens, not all mutagens are carcinogens. There are short-term mutagenicity tests which are used principally to screen chemicals for possible carcinogen risk and thus move them into more sensitive procedures. Tests involve placing pesticide on animal tissue cultured outside a living animal, or placing the pesticide in a living animal. In both methods, samples are taken and examined for signs of chromosome damage from mutations. Another test is designed to see whether animals activate mutagens or deactivate them in their bodies. This is the 'dominant lethal assay', and requires a microorganism to be injected into an animal while a dose of the pesticide is also given, but not by injection. After a while the microorganism is removed from the animal and examined to see if there are more or fewer mutations than when it is treated with the pesticide on its own, outside the animal.

Reproductive effects

Looking for chronic reproductive effects is more difficult because reproduction can be interfered with in so many ways. For example, males can be affected by a reduced fertilising ability of their sperm. In females, there can be reduced implantation of eggs in the uterus, disrupted foetal development, and interference with placental function. Some of these effects may not be evident until a reproductive cycle begins, thus making them difficult to discover in acute and even chronic toxicity studies. Nor need the effects appear in the exposed animals, but perhaps only in their offspring, and then it could be one or two generations later.

Teratogens produce birth defects because of their influence on the developing embryo. Even here the effects are not simple. Chemicals may kill an embryo, cause it to be reabsorbed into the uterine lining, or deform it. The effect depends not only on the chemical but also when it is introduced.

Tests used to detect teratogens include watching up to three generations of an animal, while observing fertility and any defects in the offspring. Pregnant animals are exposed to pesticides just when the embryo's organs are developing to see if deformities are induced. Like the other tests for longer-term effects discussed above, these tests are not very reliable and results are extremely difficult to apply to humans.

Toxicity to humans

The main data for making judgements on human toxicity are extrapolations from animal data and epidemiological data. This is essential because of the ethical barriers to conducting suitable tests on humans before pesticides are released onto the market.

Applying animal data to humans

Even though the use of animals, insects, and even bacteria in laboratories gives far greater control over the experimental conditions than is possible in the real world of humans, the data gained are of limited use. The differences among species in their responses to pesticides make it difficult to argue for any direct relationship between test animals and humans. One toxicologist observed that 'man is not just a large rat'. Indeed, it is often argued that the pig is one of the closest animals to humans as far as metabolism and internal organs are concerned, and therefore it is interesting to note that it is one animal *not used* to test pesticides for safety.

Extrapolating data from animal cancer tests to humans is fraught with problems. Many scientists believe there is no such thing as a safe dose of a carcinogen, since one molecule may be enough to initiate a cancer. This is the basis of the Delaney clause in the US Food, Drug, and Cosmetic Act. This clause stipulates that any substance found to produce cancer in an animal under test conditions should not be allowed in human foods. The powerful Regulatory Council of the US government concluded in 1979 that 'because there is no currently recognised method of determining a no-effect level for a carcinogen in an exposed population, substances identified as carcinogens will be considered capable of causing or contributing to the development of cancer even at the lowest doses of exposure.'[10]

It is easy to see why, in the UK, both industry and government regulators are wary of acknowledging that anything less than the 'no observable effect level' might be harmful. Were they to accept that possibility, then the minutest concentration of a carcinogen would have to be accepted as presenting *some* health risk. Because analytical techniques are improving all the time, carcinogenic substances can now be detected at the minutest of concentrations like parts per billion and, with tough criteria like the Delaney clause in force, pesticides which may have been approved for years may soon be removed from the food supply in the US.

There are other difficulties. Scientists cannot be sure that test animals and humans react in the same way to carcinogenic chemicals. They do know that different animals react quite differently, some being more sensitive to a given chemical and others less sensitive. Laboratory test animals are selected so every animal is as near identical as possible, and they are meant to be healthy and are kept healthy. Humans are a more varied bunch, as is dramatically evident in the differences in the faces on the street. Few of us humans are healthy all the time, and many are unhealthy much of the time. General health affects how the body's defence mechanisms cope with an initiated cancer and thus affects the number of initiated cancers which develop to damage health.

None other than the US National Cancer Institute summed up the dilemma in due scientific language: 'Although an attractive idea, quantitative risk assessment involving extrapolations from animal data is not yet sufficiently developed to be used as a primary basis for regulating human exposure to carcinogens. . . . quantitative extrapolations involve potentially large errors, some of which could underestimate the actual human risk from exposure. Scientific knowledge is currently insufficient to lend precision to this process.'[11]

Concern is as strong in the UK. Evidence given to the House of Commons Committee on Pesticides and Health[12] was clear about the limitations of animal tests. The Advisory Committee on Pesticides, which is responsible within the MAFF for assessing the animal tests and recommending approval or not, admitted there was no 'equivalent test to giving compounds to man'. Another witness said, 'the methods of assessing toxicity in animals are largely empirical and unvalidated,' and one claimed, 'when the methodology of conventional test procedures is discussed it becomes evident that quite small differences in procedure can have large effects on outcome and so on the inferences that can be drawn from them'. The Committee recommended that 'an evaluation of animal testing be carried out immediately'.

The difficulties in using such questionable tests for assessing toxicity to humans were spelt out to the Committee by a representative of the British Toxicology Society. First, 'the implications for man in quantitative terms of virtually any long-term dose in animals are not known'. Second, 'there is no reliable method to extrapolate data between different laboratory species and between different laboratory species and man'. And third, 'the evidence to date would indicate that animal carcinogenicity studies are of value in detecting strong carcinogens but do not offer a simple means of discovering, nor are a reliable means of predicting, human hazard from weak carcinogens. The studies lack specificity.'[13]

These difficulties in interpreting data from animal studies can lead to difficult policy decisions on particular pesticides. However, while animals are being used to test for carcinogenicity, any positive test should cast a heavy shadow of doubt over the pesticide concerned. But, in 1989, when the UK government was defending its approval of the growth regulator daminozide (Alar), it appeared reluctant to accept the results of official tests in the US that had found cancers in mice and rats. While in the US the maker asked the government to withdraw approval for Alar, a UK government minister, Baroness Trumpington, wrote to a national newspaper, claiming that 'very high doses produced some effects in mice, and the ACP are now *considering whether or not these are significant*'[14] (my italics). Did she mean 'whether cancers produced in mice have any significance for humans' or 'whether the risk of cancer is large enough to make it worth the ACP's taking action to withdraw the pesticide'? The answer may never be known because the ACP is not obliged to explain what evidence is considered or how their conclusions are reached. However, on 13 December 1989, the junior Minister for Agriculture, David Maclean, announced that he had accepted the ACP's finding that 'there is no risk to consumers from this pesticide'.

Epidemiological studies

The second data source in studying toxicity to humans, epidemiological studies in free-living populations, has some advantages in dealing with the most direct evidence of pesticide harm, but the populations are almost completely outside the researcher's control. The environment in which humans live cannot be controlled, the necessary period of study in humans is so long, and the number of environmental factors to monitor so large, that it is all but impossible to record even the significant environmental changes that could be influencing the study population.

Populations can be studied by identifying one group which already has experienced, or will be experiencing, exposure to a pesticide. This may be the accidental exposure of workers in a factory or farmers during their normal work, or the deliberate exposure of volunteers who agree to be exposed to a test chemical in an experiment. There must be a matched group that is equal in all respects to the other group, except for the exposure to the pesticide. Both groups then need to be studied for perhaps 10 to 30 years, during which time the lives of the members should ideally see no changes, or at least no changes that cannot be noticed, measured and accounted for in terms of their health.

Another approach is a retrospective study. Here a group of people who have been exposed to a pesticide in recent history, perhaps in a factory, are identified, contacted and their health investigated up to the present and then monitored into the future. Again a matched group of similar age, race, occupation, lifestyle, and place of residence is needed as a comparison.

Epidemiological studies face formidable problems. Toxicologists are looking for the long-term effects of very low doses of chemicals that may produce rare effects. Thousands of people are necessary in studies designed to measure such small changes. Such studies are prohibitively expensive and get harder to control as the number of participants increases. Smaller numbers usually mean the expected changes are either unnoticed or cannot be shown statistically to be other than chance. That is, they may fail to show statistically significant results, and the information may be rejected as useless.

One of the biggest difficulties is the great variation among free-range humans in how their bodies respond to toxins like pesticides. Not only do different people respond differently to different doses, but the doses can also give different effects. Additionally, there are residues of so many pesticides in human tissues, and more in the food and water taken every day, that it is difficult to identify the causes of any long-term effects or even the chemical responsible, assuming some effects are identified. Sir Richard Doll, the British scientist who eventually proved that cigarettes caused cancer, estimated recently that perhaps 70–80% of all cancers are caused by factors in the environment, but he added that it was difficult enough to identify them, let alone prove the causal links.

Long-term effects of chronic exposure, like cancers, may not appear for 30 or more years, so any effects being observed in human populations today may be due to exposure to chemicals in the environment long ago. Also, this long latent period means that epidemiological studies cannot help in testing for possible carcinogens among the new pesticides which companies are offering for approval now. It may be 40–50 years before useful data are available, and innovative companies are unable to wait so long.

Even if the shortcomings in the epidemiological methods are tolerated, it can be very difficult collecting data to analyse. Retrospective studies of workers are often the best starting point because workers tend to have higher doses and at least some of the chemicals to which they have been exposed are known. However, few companies are willing to release data on exposure, significant changes in working conditions, or records of health and accidents among the work force. Even health records from the National Health Service are not as good as they might be. Pesticide poisoning or cancers caused by pesticide exposure do not appear in national statistics on death, which is a pity because the records are so conveniently classified by occupation, region, age, and class. One problem is that death certificates rarely indicate 'pesticide poisoning' as the cause of death. If a poisoning causes liver failure, for instance, then that failure will be entered as the cause of death. With cancer, of course, the specific cause is in any case rarely identifiable.

There *are* statistics on illness in the population based on a sample of GPs. However, since most GPs are unaware of the symptoms of pesticide poisoning and, even if they are, are usually unable to identify the source of exposure, these data do not indicate either the nature or prevalence of pesticide poisonings.

One of the most useful potential sources of epidemiological data is ignored by government. Baseline data that set out to measure important indicators of possible poisoning in carefully structured groups in the population, before any expected exposure, could provide a valuable bank of data for answering future questions about the effects of present and expected environmental contamination. There could be tissue samples for residue analysis in major groups of workers likely to be exposed to particular chemicals, as well as urine samples and more detailed residue samples of food, water, and the environment, with all the data held centrally and accessible to relevant research programmes.

Epidemiological data are contentious at the best of times because of the necessary gap between perfect scientific criteria of study designs and the imperfect data available to researchers. In addition to that are the problems of trying to study the effects of very small amounts of pesticides taken into the body over many years, as well as the frequency with which any suggestion of harm is dismissed with the claim that 'there is no evidence that pesticide X is harmful'. A common way of dismissing epidemiological data is to undermine the scientific 'validity' of the study which means the results can be ignored. This occurred when the herbicide 2,4,5-T was under scrutiny by the British government in 1979. Despite the scientists of the US Environmental Protection Agency having considered the pesticide to be sufficiently proven as a cause of miscarriage to ban it, the British line was lofty. The government dismissed the US data as insufficiently scientific and concluded that claims that 2,4,5-T was harmful were 'either not sufficiently documented or, where they have been thoroughly investigated, cannot be substantiated'.[15]

Environmental toxicity

Investigating environmental effects can be difficult because, while each chemical type may suggest some effects based on its properties, the range of possible routes of harm and of the organisms involved is impossible to predict accurately. Pesticides may affect wildlife in two main ways: first, they can directly kill non-target organisms, plants or animals; and, second, they can interfere in the food chain of non-target organisms. The environment also changes over place and time, as does the variety of species living in it. The same pesticide may cause different damage on different sites and at different times. Clearly, it is unreasonable to ask companies to cover all possible avenues for damage in their submission for an approval. They are required, as a minimum, to provide information on acute oral toxicity for at least one species of fish and one bird species other than the domestic hen. Toxicity to honeybees must be presented for any chemical likely to be applied when crops are in flower.

Additional data are indicated by the chemical type or the method of applying it. Some of these data can be gathered in laboratories, but others are found from detailed flora and fauna surveys in the field. All this means that the requirement to give an account of the environmental effects of new chemicals must be an inadequate guide to possible environmental harm once chemicals are being widely used.

Normal testing takes no account of the toxicity from inter-actions among pesticides in the environment. In normal use, these can result either from different applications of pesticides on the same site at separate times, or from mixtures of pesticides in one commercial product. Studies organised by the Institute of Terrestrial Ecology at Monks Wood looked at pesticide interactions that they thought might lead to increased toxicity. The studies were prompted by the near absence of studies on pesticide interactions. They found that when the fungicide perchloraz was given to partridges along with the insecticide malathion, there was a dramatic increase in bird deaths over that expected from the same dose of each pesticide given alone. Initial tests also show that similar interactions occur between dimethoate and chloropyriphos when perchloraz is present, and also between the insecticides carbaryl and malathion.[16]

Untested chemicals also abound in two other forms. First are the contaminants of active ingredients, otherwise known as impurities, which need not all be declared, or, when they are, little may be known about them. The highly toxic constituent of 2,4,5-T, dioxin, is a contaminant of manufacture. It is responsible for the long-term 'poisoning' of the soil, and caused the deaths after the Seveso incident in Italy when a dioxin-loaded cloud escaped in a factory accident. The second are the metabolic breakdown products of pesticides, either in the environment or in animals. While some are known for certain pesticides, the complete range of chemical interactions cannot be known in advance of the pesticides being used in each new situation. The introduction of food irradiation is a source of worry for this reason. It is very possible that pesticide residues and metabolic breakdown products could be changed by the irradiation process into new and possibly dangerous chemicals in the foods. The British government may be testing the effects of irradiation on the chemistry of foods, but who is testing the effects of irradiation on the pesticide residues in foods?

These chemical changes are difficult to identify, but the present testing processes make little systematic attempt to have them or their potential investigated, and, hence, assurances of safety carry little conviction. When the Minister for Agriculture told Parliament on 3 April 1990 that the current procedure 'ensures that only pesticides found to be safe are approved for use', he either showed too much gullibility for the job at hand or was a little economical with the inadequacies of the 'safety' testing procedures.

Dramatic effects may be easily identified, but many subtle, though no less damaging, effects may not be detected until significant consequences have occurred through widespread use of pesticides in the environment.

Inadequacy of the approval procedures

All tests have limitations. The tests used for pesticide safety do have predictive value but must never be seen as either infallible or the last word on safety. That would be a nonsense. Those who dismiss consumers' concern about pesticide risks by parading apparently conclusive tests of safety exaggerate the precision of the tests.

How the tests are conducted and controlled, as well as how their results are interpreted, have a powerful influence on the meaning of safety in the pesticide approval process.

Who conducts the safety tests?

It is not the government but the industry seeking approval that conducts the necessary tests.

In 1976/7 in the US, the largest commercial laboratory undertaking tests on behalf of pesticide manufacturers, International Biotest Laboratories, was caught falsifying test results. Something like 200 pesticides were involved, and the data had been used to justify registration around the world, including the UK. Canada found itself with 113 of these fraudulently approved pesticides, covering 'most of the major insecticides, fungicides, and weed killers used in Canada and in the production of virtually all imported foods'. These had to be retested and the data evaluated again at a cost to industry of about Can$4 million. The UK government was sheepish about how many of the chemicals approved in the UK had relied on, or even used Biotest data, claiming that approvals information was not made public. Friends of the Earth claimed there were 20 such chemicals approved in the UK and called unsuccessfully for their withdrawal until new test data were available.[17]

That such a glaring fraud surfaced is not surprising, but are other laboratories falsifying data? In the US there is now an official laboratory audit. This has revealed that deficiencies existed in many of the laboratories reviewed. In 1982, a report to the US Congress showed that 4 out of 83 laboratory audits had been referred to the Justice Department for possible criminal action.

What is perhaps more alarming is that evidence produced during the trial of two Biotest employees in 1983 showed that the pesticide-manufacturer clients of Biotest knew that data were being falsified: 'In some of the studies where final reports made claims for observations that weren't made, the clients were believed to have been well aware of the situation prior to their submitting the final report to the US Government. In at least one instance, the client [Monsanto] is believed to have been knowledgeable about the usage of reported extra animals [in the study on Machete] prior to the submission of the final report to the Environmental Protection Agency.'[18] The Congressional subcommittee conducting the investigation also reported that 'major pesticide manufacturers have built their own toxicology laboratories with a desire to gain complete control over the quality of experiments done on their products . . . Uneven quality and quality assurance programs [*sic*] persist in the toxicology testing industry.'[19] The Law Reform Commission of Canada sought to reduce the chances of any repetition with specific recommendations to government that, first, companies be required by law to notify the government of any studies or other evidence within their knowledge that suggested a pesticide was harmful, and, second, that any pesticide shown to have any invalid test data supporting its registration be withdrawn.

The British government does not check the quality of test procedures and results to the same degree. It introduced a Good Laboratory Practice code which sets out standard methods of analysis, but it does not provide for a systematic and routine checking on *how diligently* each laboratory carries out the tests.

Review of old pesticides is too slow

Even the MAFF admits that there are over 250 pesticides that were registered in the UK before 1981, when the approvals procedures were much less rigorous than today's. There is general agreement that many of these chemicals would fail current safety tests and so would be struck off the approved list. There are two problems in checking these old chemicals. First, the review process is so slow that even the MAFF says it will take an additional 30 years to clear the backlog. The second and more alarming problem for those who appreciate human nature is that the committee responsible for the review procedure is the very same committee which gives approval in the first instance.

There are pesticides on the approved list that have been in use since the 1960s. Such safety tests as were conducted then were crude by present standards. They looked for acute toxicity, but with less sophisticated tests, and paid insufficient attention, if any, to chronic harm and environmental damage. It is urgent that these pesticides be retested by the same criteria as newer pesticides. Nobody, certainly not outside the circle of manufacturers, knows the health risks from these older pesticides. There have been suggestions that all these should be withdrawn until new test data have been submitted and processed by the ACP.

The outcry from groups like Friends of the Earth and the Agriculture Committee of the House of Commons Select Committee has produced new promises of action from MAFF. In April 1990, the Minister announced increases in the staff processing data on pesticides. It is planned to increase the number of old pesticides reviewed each year from 12 in 1990/91 to 29 the following year, and to a peak of 37 per year in 1992/3. A priority system is being worked out so the most dubious will be reviewed first. The speed of this work will depend on the number of alarms over other approved pesticides. Every time a scare is reported and the ACP is asked to investigate claims of harm from a pesticide thought to be 'safe', resources are diverted from the reviews into the urgent inquiry. Urgent inquiries should be initiated, and carried out thoroughly, but there should be the resources for the reviews as well.

It is expecting too much of the ACP members to make them responsible for reviewing their earlier approvals. It has recently been proposed that all pesticides be reviewed every 10 years, though some would prefer that it were every 5. Even with a 10-year period, it would be only human for the ACP members to tend towards minimising criticism of their own previous approvals. How could any committee err on the side of safety if it meant that a significant proportion of its earlier approvals were dropped on second viewing? The more chemicals that are dropped on review, the more it may appear that earlier decisions were unsound, even though that need not be true. Clearly, new data will be presented over time as chemicals are used, and new techniques for detecting damage will emerge. Another fact is that, were the ACP's own review to lead to an excess of rejections, this might expose the underlying inadequacy of the data-gathering process and the significant element of subjective opinion in the approval process.

It would be far better if a separate organisation were established, both for monitoring pesticides in use and for conducting the regular reviews. Ten years is also a long time in the environment. Far better to have a review system where every new approval is reviewed after 5, 10, and 15 years with 10-yearly reviews thereafter. There are after all only 5 or 6 new chemicals considered for approval a year at present, though this may increase to an eventual maximum of 25 with new resources. This could be backed by continuous monitoring, with a statutory duty on manufacturers not only to carry out specified tests, but also to supply data on any adverse effects on the environment or humans they discover from any source. This would both resource the periodic reviews and supply data which would allow a special review to be initiated at any time should the monitoring suggest there were unacceptable effects of the pesticide in use.

Delaying the approval of new chemicals

New pesticides are a double-edged weapon. Some argue they are necessary to keep ahead of pest resistance and that they contain safer chemicals. Others see them as a necessary part of the commercial logic of giant chemical companies in keeping ahead of the competition while keeping farmers hooked on a chemical treadmill. Either way, the greater share of pesticides used today are the old, well-tried faithfuls, which also account for the bulk of profits in the industry. The herbicide atrazine has been around since 1957, the insecticide dimethoate since 1951, and the fungicide zineb since 1943. But many of these older chemicals, still in current use, are on the dangerous end of 'safe', and are being reviewed. New chemicals, if assessed with even tighter criteria of safety, could make it easier to drop the less friendly old standbys. The problem is getting new alternatives through the approval process. Like reviews, new applications are suffering long delays.

The Harpenden Data Evaluation Unit (DEU), which receives and assesses the data provided by companies seeking new approvals, employed only 12 scientists in 1989. The DEU passes its conclusions on to the ACP via its Scientific Subcommittee (SSC), and the ACP passes its advice on to the Ministers. In 1988 and 1989 there were 12 and 15 applications for approval, respectively, but only 5 were evaluated in 1988 and 6 in 1989. The longer manufacturers have to wait, the greater their costs and the less their interest. Not only may competitors catch up with any advances in technology, but also patents last only 20 years and the longer the delay before hitting the market, the less time before competitors can copy the product. Other countries have more rapid approvals. France gets them through in 10 months while the US takes between 1 and 2 years.

In a rare outburst of conviviality in 1989, the British Agrochemical Association (BAA) teamed up with Friends of the Earth, the Green Alliance, the Pesticides Trust, the Federation of Women's Institutes, and the Transport and General Workers' Union to lobby government Ministers. In their list of suggested improvements was a demand to reduce the time taken to consider new applications from 4 years to 1. Following the collective lobby, the Minister of Agriculture announced the allocation of new resources to the DEU, an action which he expected would speed up evaluations. He anticipated a tripling of capacity, with 8 new chemicals to be considered in 1990/91, 12 the next year, and 20 in 1992/93. He hoped that by 1994 there would be capacity for evaluating 25 new chemicals a year.

These are still only expectations, and it remains to be seen whether the new toxicologists can be attracted to the DEU. But resources are not the only problem. With more stringent demands on the criteria of safety and the clear opportunity to require not only better data but more environmental studies, the greater amount of data supplied may take longer to evaluate. There have already been claims that some of the data supplied for new products have been substandard and have had to be returned. This increases the time taken for approvals.

It is important not to assume, however, that new chemicals are automatically better. A critical review of MAFF's approval arrangements in *Farmers Weekly* quoted a toxicologist involved as saying, 'Only one or two new products stand out as better chemicals during the past two years. When you look at the track record of compounds going to the Scientific Subcommittee, a regrettable number either have poor data packages or the pesticides themselves are not safe. By safe, I mean a better and more effective chemical that has a clean, long-term toxicity package and is not mobile or persistent in the environment.'[20]

REFERENCES

1. FoE (1988). *An investigation of pesticide pollution in drinking water in England and Wales*. A Friends of the Earth report on the breaches of the EEC Drinking Water Directive. London: FoE.
2. Wilkinson, C. F. (1987). Environmental toxicology; its role in crop protection. In *Rational pesticide use*, eds R. K. Atkin and K. J. Brent, Cambridge University Press.
3. MAFF (1990). Press release No. 45/90. London: MAFF.
4. NRC (1980). *Regulating pesticides: a report prepared by the committee on prototype explicit analyses for pesticides*. Washington, DC: National Academy of Sciences.

5. Law Reform Commission of Canada (1989). *Pesticides in Canada: an examination of federal law and policy*, p. 53. Ottawa: Law Reform Commission of Canada.
6. Robbins, C. J. and Bowman, J. (1983). Nutrition and agricultural policy in the United Kingdom. In *Nutrition in the community*, ed. D. McLaren, pp. 183–200. Chichester: John Wiley & Sons.
7. Boardman, R. (1986). *Pesticides in world agriculture: politics of international regulation*, p. 82. London: Macmillan.
8. Rose, C. (1986). Pesticide controls: a history of perfidy. In *Green Britain or industrial wasteland?*, eds E. Goldsmith and N. Hildyard. London: Polity Press.
9. MAFF (1986). *Data requirements for approval under the control of pesticides regulations, 1986*. London: MAFF.
10. US Regulatory Council (1979). *Regulation of chemical carcinogens*, p. 6. Washington, DC: US GPO.
11. NCI (1979). NCI draft memorandum to FDA on use of animal data in cancer risk assessment. 8 Chemical Regulation Reporter 274, p. 275. Quoted in Law Reform Commission, op. cit, p. 56.
12. House of Commons (1987). *The effects of pesticides on human health*. Vol. 1. Report and Proceedings of the Agriculture Committee. Second Special Report, Session 1986–7, p. xv. London: HMSO.
13. House of Commons. Op. cit., p. xvi.
14. Trumpington, Baroness (1989). Letters page, the *Guardian*, 6 November.
15. Crocker, J. (1979). 2,4,5-T: contradictions continue. *Chem and Ind*, p. 803, December.
16. Dawson, A. S. (1990). Interactive effects of pesticides with partridges. In *Institute of Terrestrial Ecology annual report 1988–9*. Swindon: NERC.
17. House of Commons. Op. cit.
18. Law Reform Commission. Op. cit.
19. US House of Representatives (1982). *EPA pesticide regulatory programme study*. Hearing before the subcommittee on department operations, research, and foreign agriculture, 97th Cong., second Session, 17 December, p. 209.
20. Harvey, J. and Wilson, R. (1990). Pile-up in pesticide programme. *Farmers Weekly*, 23 February, pp. 73–6.

5 The Politics of Pesticide Approval

★ Science produces opinions about facts but absolute certainty about nothing.

★ With pesticides, the same science which used to 'prove' safety is often rejected as a 'proof' of harm.

★ The manufacturers of pesticides conduct the safety tests which are used to decide whether approvals are granted for their products.

★ Pesticide approval and regulation is highly political in the UK and internationally.

★ Consumer confidence in pesticide regulation requires less secrecy, and greater involvement of consumers in all decisions.

Science the equivocator

Science, of course, can never provide absolute certainty. The best it can offer is a reasonably objective way of testing ideas and theories put up to explain things that go on in the real world. The scientific method is designed to reduce the influence of scientists' personal opinions, hunches and deceits to a minimum and to offer a test that can be repeated by any other scientist, anywhere, thus providing a check on the results. When a scientist says something is 'proved', all that is meant is that the likelihood of its being true is high, or better than chance. Thus, 'scientific fact' is a misleading term for something which is only 'most likely to be true'.

In practice, the method is often poorly followed and many experiments are flawed. More common still are wide differences of opinion about the results of any experiment. Just consider the arguments raging during the past 40 years over diet and heart disease, despite its being the most thoroughly researched medical subject, when judged on the number of published papers. This doesn't mean that science provides false information, merely that opinions about experimental data are as important as so-called 'scientific facts'.

Pesticide testing should follow the most stringent scientific methods available. However, as Chapter 4 shows, the reality of testing is very different. So, despite all the waving around of safety tests by both industry and government, it is opinions, not facts, that decide whether a pesticide obtains approval and whether demands for a withdrawal are justified.

Yet scientists and scientific 'facts' are paraded whenever pesticide safety is discussed, even though, and contrary to society's general belief in the omnipotence of science, it is rare for any facts or conclusions it produces to be beyond dispute. More often they *create* disputes.

When the trade unions took up the cudgels to fight for the safety of forestry and agricultural workers who were exposed to large amounts of 2,4,5-T, they were treated with derision by the regulatory authorities in the UK. Their dossier of cases covering UK workers' families, and members of the general public affected after exposure to the chemical, was dismissed as being insufficient evidence, despite the government's making no attempt to collect 'sufficient' evidence of its own.[1] The Catch-22 of science in the service of humankind was brought into play to block the attack on pesticides.

The authorities dragoon 'science' into defending their position that pesticides are safe. It is sufficient, apparently, only for the Minister to say, 'the Advisory Committee on Pesticides has concluded . . .' for safety to be 'proven'. However, when science is used to show that a pesticide is harmful, it is dismissed by the government as being a limp weapon. Their logic is that scientific claims cannot be 'beyond reasonable doubt' or even that the data 'are not conclusive'. There is, apparently, science that confirms the safety of pesticides but no science reliable enough to confirm their harmful effects. Science is being used to hinder and hood-wink when it should be informing and clarifying decisions.

Perhaps the greatest flaw in the government's defence of pesticide safety is that they themselves are not being 'scientific'. Industry-supplied test data are never submitted to peer review, the well-established process that weeds out most bad science before it is seen by other scientists – which is long before it reaches public attention. It operates through a system whereby scientific claims make their first appearance in recognised journals. Before a submitted paper is published, the editor sends it round a select group of referees, who comment on methods used, statistical analysis, originality, and any missing data needed to justify conclusions. Comments are sent to the author, who must make adjustments and resubmit. Only when modified to the editor's satisfaction is the paper published. The purpose of this big event is not merely so that other scientists can be entertained. It allows the paper to be criticised and challenged by peers and is crucial to the validation of scientific work. At present, no similar vetting procedure is attempted with the results of pesticide safety tests. They are done, and accepted, as fact.

Science and the consumer

While it is true that scientific achievements are often awe-inspiring, non-scientists should not be conned into feeling inadequate to the task of judging the implications of scientific findings. We should not be blinded by the mysteries of computer technology or light-flashing laboratory apparatus. These are mere tools and may be no better understood by an operator than is a washing machine or microwave oven by a householder.

It is the conclusions drawn from experiments and from discussions about data, not the elaborate technology, which are of use to society. There is every reason to believe that non-scientists are as equipped as scientists to form valid opinions on the implications of these conclusions. They may even be *better* judges since they are free from scientists' vested interests. Currently, however, a major handicap for consumers is that we are excluded from access to the relevant data.

When politics rules safety

Science is certainly not perfect, but at least honest scientists can discuss the imperfections in their data and the conclusions each of them might be drawn to. It is interesting that the decisions on whether or not to approve a pesticide are taken not by these 'experts', the scientists and toxicologists who understand the data, but by Ministers or administrators. The Advisory Committee on Pesticides and its supporting Scientific Subcommittees merely give advice and recommendations to the Minister and the government. Any decisions are made jointly by the Ministers of Agriculture, Employment, Health, and Environment, together with the respective Secretaries of State for Scotland and Wales. This immediately puts pesticide approvals into the political orbit.

Politics is not only about government and political parties. There is also a wider meaning which is best seen in the original name for economics – political economy. This takes account of power in decisions about how society's resources are used or distributed. Power is nothing so blunt as physical force. It operates through more common, more subtle channels, which are legitimate and involve us all, even though individually our power may be negligible. Real power resides in wealth, bureaucratic processes, rank, access to information, privilege and even personality. The politics of pesticides is about the influence of vested interests in the regulatory process.

The Ministers' decisions are as much to do with economics and lobby power as with science. Hence, when a decision about an approval or a withdrawal is announced, the real reasons behind it are unlikely to be stated. The political orbit runs from vested interests in the production and use of pesticides, through the MAFF, and on to the highest of international bodies like the World Health Organization.

In the UK, the old voluntary Pesticide Safety Precautions Scheme (PSPS), which 'relied on extensive and close working contacts between government officials and the agrochemicals industry', excluded all participation by other members of society. There was no public access to the criteria used in decisions on specific pesticides. After the UK joined the EC, products legally approved under EC statutory rules but not agreed under the UK 'gentleman's agreement' were allowed to be imported, thus threatening the cosy market of UK manufacturers. So the UK industry, through the British Agrochemical Association (BAA), tried to convince the government that the voluntary scheme should be used to produce a restricted list which could be used to block imports from other EC countries, even though their products had the backing of a legal registration system. Furthermore, the UK was bound by the Treaty of Rome to let them in.[2] Since the biggest pesticide manufacturers and retailers in the UK are either subsidiaries of multinational companies, or are wholly owned by them (e.g. Shell, Du Pont, Bayer, Hoechst, Monsanto), they often have an interest in creating and maintaining trade barriers between countries, even countries where they have subsidiaries. The reason is simple. It keeps the world market divided into parcels, and there is nothing like keeping each local company protected from outside competition.

When the trade unions launched their campaign against 2,4,5-T, they argued for a tighter system to regulate pesticides. The Health and Safety Executive was proposed as a Watchdog body because it already had regulatory powers over the use of chemicals in the workplace and worker representatives involved in decisions. This presented another threat to the 'gentleman's agreement' schemes, and 'the room for manoeuvre of the newer bodies was constrained by the continuing importance of the traditional regulatory bodies, and by the government–industry relationship built up over many years through PSPS consultations.'[3] The determined stance of the pesticides industry was displayed by the then chairman of the BAA, Dr David Hessayon, also author of the popular 'Expert' series of gardening books. He said, 'If we give way on 2,4,5-T, the unions will then go on to campaign against another chemical and then another. We are engaged in a power struggle for the control over the industry which we cannot afford to lose.'[4] Surely, if the safety tests were adequate and conclusive all such campaigns would have been defeated easily?

Risk assessment

When approving a new chemical, and especially when considering withdrawing one, there is usually an attempt to weigh up
benefits against costs or risks. Risk is an elusive but useful
concept. Everything has some risk be it of negligible or significant
consequence.

Risk is really chance. On tossing a coin, there is a 50:50 chance
of getting heads. This is the same as saying the risk of getting
heads is 1 in 2. Some risks may be greater than others in mathematical terms, but whether they are more or less acceptable to
people depends not on the mathematical probability, but on
individual preference. Driving up the M1 carries a small risk of a
fatal car accident, which few drivers would choose. However,
motor racing is a sport based on the *thrill* of taking that risk.

So who should decide when the risks from a pesticide are
important? When the Minister says 'there is no risk of harm' for
some pesticide or another, what really should be said is, 'We have
estimated the risk of harm but don't think enough people would
be made ill or die to cause a stir.' Consumers should at least be
given the chance to decide whether they want to take on particular risks. Too often it is said that we common consumers are
unable to handle such complexities as risk, but a brief visit to any
one of the several betting shops on any high street, on any week
day will give a realistic picture of willingness to deal with risk, if
not skill. Risk assessment, even if not undertaken consciously or
with the aid of a computer, is part of everybody's daily decision-
making.

There are formal methods of estimating risk and weighing it
against benefits as an aid to decision-making. Such procedures
are rarely used, and certainly are not required in the statutory
regulations governing pesticide approval in the UK, Canada, the
US, or the EC. Such risk assessment cannot be reliable because
too many assumptions have to be made. The safety tests have
been shown to be at best equivocal (see Chapter 4). There are
additional problems in assessing the possible risk to humans or
the environment from harm that may not appear for decades and
which is usually impossible to predict. Also, should voluntary
and involuntary exposure be treated differently in assessing

risks? And how should we treat different *levels* of exposure? Workers applying chemicals could be more at risk than the general public; rural dwellers are more at risk from spray drift than the urban population. Different cultural groupings and simple diet preferences could make for widely uneven distribution of food residues and hence exposure to different risks.

Even the benefits of pesticides are difficult to measure. The BAA lists nine in a promotional leaflet, *Why Use Pesticides?*[5]

- improving food quality
- reducing food prices
- maintaining public health
- promoting animal welfare
- banishing drudgery
- aiding habitat management
- helping gardeners and groundsmen
- controlling aquatic weeds
- earning export revenue

All of these are open to alternative views on both the role of pesticides and the size of their contribution. Improving food quality? The BAA leaflet includes unblemished fruit and vegetables, but denies any 'difference in taste, nutritional value or "healthiness" [*sic*] between conventionally grown and organic food'. Reducing food prices? The leaflet lists the examples of potatoes in 1985 being 45% cheaper than in 1975, and of bread being 8.5% cheaper, with the assertion that pesticides 'have made a major contribution to this satisfactory situation'. In fact, given the contributions from new varieties, fertilisers, reductions in machinery and total labour costs, and productivity gains in bread making, plus the general price-lowering effects of competition from other foods, the specific role of pesticides could be small and difficult to get any two economists sitting side-by-side to agree upon.

The public is left out of these discussions of risk and benefit and the louder, better organised and legally entitled voices of pesticide manufacturers, civil servants and their appointed advisers are the only ones given votes around the decision table.

The role of international regulation

The international clearing house for toxicity is the joint Food and Agriculture Organization (FAO) and a WHO body called the Codex Alimentarius Commission (CAC). It is primarily concerned with international food standards, and, because of the amount of trade in food, regulating pesticide residues in foods has become an obvious and important activity. The CAC, now the key body in the international regulation of food residues, has a subgroup on pesticides, the Codex Committee on Pesticide Residues (CCPR). It devises common standards for residue analysis and sets acceptable levels of residues in food and feeding stuffs. The work is highly political inasmuch as acceptable limits of residues are, and have long been, used as non-tariff trade barriers between nations.

Representatives on the CAC are nominated by member governments, and there are also expert advisory meetings, called Joint Meetings, between FAO and WHO specialists. These meetings serve to advise the Pesticide Committee of the CAC. The decisions on limits are taken by the representatives of the member governments on the CAC, indicating the high level of political pressure behind every decision. At the heart of the process are representatives of the pesticides industry along with a caucus of voluntary organisations.

The whole process operates on the assumption that pesticides *will* be used and that some risk is not merely tolerable but inevitable. It is a venue for reaching compromise among all the international interests in pesticides, including health, environmental and commercial. Reliance on fact or science has been less important than reaching agreement. Pesticides become the object of protracted negotiations where the data depend on what the member governments and interested companies choose to make available. The result is serious gaps in data and great variation in their quality, depending on what is at stake.

There are several opportunities for industry to exert its influence. It often provides formal members of national delegations to the CAC or CCPR. In 1982, the UK delegation of nine included government officials plus one person from the then Food and Drinks Industries Council (now the Food and Drink Federation) and one from the BAA. Such conviviality is normal. The international pesticides lobby group GIFAP (International Group of the National Associations of Agrochemical Manufacturers) has observer status on all CCPR meetings. From this position representatives make statements to the Committee that are published in the proceedings and 'can have an appreciable impact on the atmosphere of the discussions'.[6]

This international activity initiated the UK move to introduce a statutory registration scheme and maximum residue levels (MRLs), both of which already existed in the EC when the UK joined. The EC developed MRLs not for health reasons, but to stop member states using their own limits to block agricultural imports from other member states. These MRLs had nothing to do with health, but it is worth noting that as early as 1978 the EC drafted, and had accepted by the Ministers, a series of guidelines on health measures including domestic use, storage, and reducing contamination of human foods. Because certain countries had already introduced bans on some environmentally hazardous pesticides, the EC also introduced an EC-wide list of banned pesticides. Again this move was aimed at stopping bans being used as non-tariff trade barriers within the EC. The best confirmation of this reasoning lay in the total failure of the EC to consider prohibiting the export of chemicals banned in the EC to third countries in the initial directive of 1976. This is still being fought for in the EC.

Secrecy and trust

Data on pesticide approvals are withheld from public gaze, and both full details of ACP decisions and their assessment of data submitted by manufacturers are 'official secrets'. This does little for public confidence in Ministers' bland and repetitive assurances.

True, not all consumers are capable of understanding the data, and probably still fewer want to. However, there are well-qualified independent researchers, like Alaistair Hay at Leeds University, and organisations, like the National Consumer Council or Friends of the Earth, with both sufficient expertise and genuine public interest, who could understand and interpret the data for lay people. The overriding justification for providing free access to safety data is simple: almost every consumer is an involuntary consumer, a passive recipient of pesticides. The involuntary consumers of pesticides have a right to know where they are used and a right to decide which chemicals they will accept being exposed to.

Public confidence in government regulation

Public confidence in government's ability to minimise risks from pesticides is low. The Agriculture Committee of the House of Commons put this somewhat bluntly in its final recommendations to the government. Recommendation 17 said:

> Having studied the detailed memoranda of the Ministry of Agriculture and heard oral evidence from its officials twice, we have concluded that anxiety can no longer be allayed by merely saying that no harmful effects have been observed in certain pesticides and that therefore they are safe. Those responsible for their clearance must convince the public that they have the resources, knowledge, and independence of judgement to investigate potential risk to human health from pesticide use and they must do this in a more open way.[7]

Full access to the day-to-day data and assessments of both new and reviewed pesticides would be helpful. But greater confidence requires a more direct presentation of safety issues without the camouflage hitherto used by Ministers to protect vested interests. For example, when 2,4,5-T was the subject of such controversy in the UK in the 1970s, the ACP could say only that there was no evidence of harm. However, in the US where the chemical was also under attack, it had been banned from use on all crops except rice in 1970 and its registration with the Environmental Protection Agency (EPA) for use in forestry and on rights of way was withdrawn in 1979. So strong was the EPA case that the manufacturer of 2,4,5-T, Dow Chemical, chose to stop production. In the UK, the ACP refused to publish the details of how its assessment of the evidence had led to the decision to continue its use in the

UK. More recently, Alar (daminozide), which was banned in New Zealand and then withdrawn in the US as a possible carcinogen, was allowed to remain on the UK approved list after a special review in 1989. The manufacturer, Uniroyal, accepted that the results of cancer studies in the US undermined Alar's registration with the EPA and voluntarily stopped worldwide sales. Neither details of how the ACP assessed the US data on cancer risk, nor any explanation of the claim that the pesticide 'did not pose a health risk to consumers' was made available. Likewise with the fungicides zineb, maneb and mancozeb, which the US EPA was considering banning in 1989 because of evidence of their carcinogenicity, the ACP failed to explain the details of either their assessment of the new data or their reason for leaving all three on the UK approved list (see pages 66–7).

Both the US and Canada have developed a more open system than the UK. The EPA officials who conduct the approvals in the US and the companies submitting data realise the contribution they are making to public confidence by submitting health and safety data for public scrutiny. Officials feel it is an essential part of their function, and industry accepts a clear duty. Even the National Agricultural Chemical Association (NACA) sees safety as the main priority and accepts the provision of information as a 'necessary part of its guarantee'.[8] The only limitation on data availability is to restrict public access to efficiency data which might be of value to competitors.

This healthy approach is also guaranteed by the Freedom of Information Act in the US, and is encouraged by strong patent laws which protect the legitimate rights of companies to ownership of chemical formulae and production processes. There seems to be no good reason why a similar freedom of information should not be provided in the UK. Many of the companies producing or marketing in the UK either do so in the US as well or have had the same pesticides registered there, where the safety data are available to the public. Some of the data presented for approvals in the UK have been both generated by the US and used in their approvals process. It is anomalous that safety data which is withheld in the UK can be obtained freely in the US. The main difference appears to lie in attitudes and not in law or logic.

As the 1985 Food and Environment Protection Act promised,

there have been some concessions on openness. Evaluation statements containing some information on new pesticides are published in the Scottish and English Gazettes when full approvals are granted. The full evaluation package, but not the raw data submitted by the companies, is available on a 'reading room' basis, but only where a scientific case is made, and approved, for seeing the data for research purposes. The quaint term 'reading room' is, however, somewhat restrictive, as no photocopies can be made though note-taking is allowed. Access to these data is tightly controlled and it is an offence to publish any information gleaned from the reading. The data are available, to use the term loosely, in the Data Evaluation Unit in Harpenden. 'Public access' is, in fact, a misnomer since the public cannot see the raw data, and individuals who do so are forbidden by law to give information to the public.

Conclusions

Despite all the skills of well-drafted, Ministerial public-relations statements defending testing procedures and the decision to keep all government data secret, every consumer knows the emperor has no clothes. Pesticides are political. They are also of increasing concern to consumers, who are rightfully demanding to be involved in that political process.

REFERENCES

1. Kaufman, C. (1986). 2,4,5-T; Britain out on a limb. In *Green Britain or industrial wasteland?*, eds E. Goldsmith and N. Hildyard. Cambridge: Polity Press.
2. Boardman, R. (1986). *Pesticides in world agriculture: politics of international regulation*, p. 83. London: Macmillan.
3. Ibid., p. 84.
4. Kaufman, C. Op. cit.
5. BAA (undated). *Pesticides in perspective. Why use pesticides?* Peterborough: British Agrochemicals Association.
6. Boardman, R. Op. cit., chapter 5.
7. House of Commons (1987). *The effects of pesticides on human health.* Vol. 1. Report and Proceedings of the Agriculture Committee. Second Special Report, Session 1986–7, para. 250. London: HMSO.
8. Ibid., para. 184.

6 Legislation and Regulations

> ★ New legislation in the UK could pave the way to tighter controls on pesticides.
>
> ★ Only 64 out of 400 approved pesticides now have Maximum Residue Levels defined in law for certain foods.
>
> ★ The Food Safety Act could be used by consumers to challenge the presence of any pesticide residue in food.

There is a long history of UK law covering pesticides. This may seem odd, given the scope and volume of campaigns for better protection of consumers and the environment. But laws are only as good as their content and the way in which they are enforced. Until the mid-1980s, UK laws deserved criticism. Significant changes have been made in and since the Food and Environment Protection Act, 1985, bringing many aspects of UK law into line with more progressive EC law and removing important deficiencies. The law is far from perfect, but as well as bringing such concrete changes as introducing compulsory training for workers handling pesticides, Ministers now have greater powers to make further improvements without resorting to the parliamentary processes of making new laws.

This chapter is not intended as a quick step to legal expertise but as a guide to what laws exist and what they cover. Anyone interested in investigating possible abuses of pesticides, taking action, or even pressing for improved protection is recommended to get copies of the relevant legislation. As legal documents, they require a little more concentration than the TV listings but are not as difficult as many people fear. There is always advice and support available from the sponsoring government department and bodies like trade unions, the National Consumer Council, and environmental groups like Friends of the Earth.

A list of the important legislation provides a brief overview of what is covered and where.

The Food and Environment Protection Act (FEPA), 1985, Part III

All current UK legislation on pesticides now comes under this Act. Its professed aim is to 'protect the health of human beings, creatures and plants; to safeguard the environment and to secure safe, efficient and humane methods of controlling pests'. Under the Act, a pesticide is defined as 'any substance, preparation or organism prepared or used, among other uses, to protect plants or wood or other plant products from harmful organisms; to regulate the growth of plants; to give protection against harmful creatures; or to render such creatures harmless'.

Not all pesticides are covered by the Act and the attendant regulations. For example, pesticides used in house paint, textiles and paper are exempt, as are the pesticides that are applied directly to farm animals like dips and wormers, which are covered by the Medicines Act, 1968. Also exempt are all pesticides used as part of manufacturing processes, such as the pesticides used on leather, in chipboard, etc. Pesticides used in or around factories, like herbicides and rat poisons, are included.

Ministers are given wide powers under the Act to:

- make information available to the public
- control the sale, import, distribution, storage, use, advertisement and labelling of pesticides
- seize and dispose of pesticides where there are breaches of regulations
- set maximum residue levels
- issue codes of practice
- empower enforcement officers

- establish the Advisory Committee on Pesticides as a statutory body

Under this Act came the:

Control of Pesticides Regulations, 1986

These regulations come under the Act and spell out the requirement for having pesticides approved before sale. Essentially, the regulations lay down two rules. First, that no pesticides can be sold, stored or used commercially without formal approval; and, second, that everyone handling and applying pesticides must stick to the conditions of the approval which in this context means the label must be followed.

Anyone who has had experience of the sometimes carefree manner in which pesticides are used on farms or in public places might appreciate the general requirement on all users of pesticides that they 'shall take all reasonable precautions to protect the health of human beings, creatures and plants, to safeguard the environment and in particular to avoid pollution of water'.

There are now three types of approval:

Experimental permits

These are granted normally for one year so that pesticides can be properly tested on site before application for approval allowing them to be sold. The permit does not cover uses beyond agreed experimental purposes, and any crop produced cannot be sold and must be disposed of in an approved way.

Provisional approval

This is for a limited time only and may be given to products not showing high acute toxicity but for which additional data need to be submitted in order that a full assessment of their safety can be made.

Full approval

When the ACP is satisfied, approval is given to sell a product for an unlimited period but this approval can be modified or revoked if the Ministers have reason to question human or environmental safety, etc. In making an approval, the Ministers specify the approved uses of a pesticide and list any conditions of its use. These are carried on the product labels, which is another good reason for reading the labels of any pesticide products. Labels should include the following information:

- field of use (forestry, horticulture, home, etc.)
- crops, plants, or surfaces on which it may be used
- maximum application rate
- maximum number of treatments
- latest time of application
- any limits to area or quantity allowed to be treated
- user-protection advice or required training before use
- advice for environmental protection (dangerous to fish, bees, etc.)

Off-label approvals

Users of pesticides may apply for approval to use an approved pesticide for purposes not already specified in the approvals, or to have the conditions of use changed. These approvals are called off-label for the obvious reason that they are additional to the limitations printed on product labels. They arose because of the tendency of manufacturers to seek approvals only for those crops they thought would provide an economic market for the products. After all, they had to carry out tests on all crops for which they wanted approvals, and tests take time and cost money. So that growers of minor crops could still use pesticides which may be efficient and safe on their crops, the scheme was devised to extend pesticide approved uses where appropriate. Because of the large number of such requests that arrived, the MAFF have drawn up a list of so-called minor uses which can be followed without additional approval from the MAFF.

Minor uses cover three categories of crop:

Non-edible crops and plants Any pesticide provisionally or fully approved for use on any growing crop can be used on crops like hardy ornamental nursery stock, forest nursery crops, and flax grown for fibre.

Nursery fruit crops Pesticides provisionally or fully approved for use on any crop for human or animal consumption can be used on nursery fruit trees, bushes and non-fruiting strawberry plants on condition that any fruit harvested within one year of treatment must be destroyed.

Crops used wholly or partly for consumption by humans or livestock This list is more restricted, but includes numerous crops on which pesticides already approved for other specific crops may be used, but with certain conditions attached. For example, beetroot and celeriac may be sprayed with chemicals fully or provisionally approved for carrots, while peppers (capsicums) and aubergines can be sprayed with products cleared for use on tomatoes. Crops like poppy, mustard, and evening primrose may be treated with chemicals approved for oilseed rape, with the attached condition that they must not be applied at flowering time if they are hazardous to bees.

Consents

The Regulations also give Consents in which Ministers set conditions on activities rather than pesticides. There is a Consent covering the aerial application of pesticides. This is detailed and sets out not only the list of pesticides that can be used but also requirements for training, consultation and record keeping. It also sets out the meteorological conditions for aerial spraying and even covers flying technique. Not all problems will necessarily be solved, however. Consider flying technique. Pilots must not fly, when 'engaged in an aerial application', less than 200 feet above a building such as houses, not less than 200 feet horizontally from a children's playground (100 feet if the owner gives permission!), and not less than 200 feet above a public highway other than a motorway, for which the minimum height is 250 feet. Just to make sure there is no spray drift with such proximity, the next paragraph in the Consent requires the pilot to 'confine the application of the pesticide to the land intended to be treated'. So there is no need to worry after all.

Another important Consent covers the use of pesticides. It requires that employers ensure that anyone using pesticides has proper instruction and the necessary training and guidance to be a 'competent' user. Also that they have the necessary certificate of 'competence'. It also bans the mixing of any two anticholinesterase pesticides unless the label of one product says the mixture can be made.

Codes of practice

In addition to Consents are codes of practice that guide users on how to take reasonable precautions to protect health, etc. There was a Code of Practice on the Agricultural and Horticultural use of Pesticides that is now combined with the proposed Code of Practice on Exposure to Pesticides at Work under COSHH (see page 110). The single code is known as the Code of Practice for the Safe Use of Pesticides on Farms and Holdings.[1] This guides on training, recognising when a pesticide is necessary and protecting workers. It also identifies recognised certificates of competence and gives advice on how to use pesticides, including choosing correct spray equipment, and on record-keeping.

Certificates of competence are available from courses approved by BASIS, the British Agrochemicals Standards Inspection Scheme. They offer courses for store keepers, salesmen, and sales assistants. Many BASIS courses are run in County Agricultural Colleges, but approved alternatives are available. Users, which include farm workers applying sprays, must have certificates of competence, which are arranged through tests set by National Proficiency Tests Council branches around the UK.

The codes are like the Highway Code in that non-observance does not constitute an offence, but in the event of a poisoning accident or some environmental damage, breaches of the codes may be used as evidence in a prosecution.

The Pesticides (Maximum Residue Levels in Food) Regulations, 1988

These were made under Part III of FEPA and are new controls for the UK. The introduction of Maximum Residue Levels (MRLs) was required in EC Council Directives of 1986. It is an offence to cause, or leave, any one of given levels of residues of about 64 pesticides in a list of specified foods. The list relates to what are called primary foods, like wheat, potatoes and milk, and not to their manufactured products like pie crust, potato crisps, or milk chocolate. The MRL legislation is aimed mainly at producers of animal and vegetable crops rather than at manufacturers and retailers. The foods covered include meat and milk products, cereals, and specified fruits and vegetables, including oranges, apples, bananas, tomatoes, cabbage, lettuce, and potatoes. When the Regulations were first published in draft form early in 1988, they listed 62 pesticides. When finally passed by Parliament, several had been dropped, including dimethoate, iprodione, the dithiocarbamates, and tecnazene although these were either established as harmful chemicals or were known to regularly exceed the proposed MRL limits. (See page 122.) No explanations were given.

The list in the Regulations of pesticides with MRLs is only a fraction of the total number of pesticides approved for use in the UK.

It is important to note that the MRL is not meant to be interpreted as a level which, if exceeded, can be taken as harmful. Although health risks are taken into account in setting the MRL, it is meant to be used specifically as a level of residue that should not be exceeded in a crop if the pesticide has been used according to approved agricultural practice. This, of course, does not mean that the MRL is a safety level. It merely refers to approved methods of use. (See page 121.)

A measure of residues that is more directly linked to health is the ADI or Acceptable Daily Intake level. The ADI is usually set at one-hundredth the recognised 'no effect level' of intake. It allows for an assumed safety factor of 100 because humans are judged ten times as sensitive as any other species and the most sensitive human is ten times as sensitive as the average human. Basically, this is guesswork. It is used to suggest that the ADI is the total level of intake that could be taken every day for life without any effect. This presents serious shortcomings which are discussed on pages 119–20.

COSHH – The Control of Substances Hazardous to Health

These new regulations under the Health and Safety at Work Act, 1974, came into force in October 1989. COSHH requires all employers to assess any risks of exposure to hazardous substances in the place of work. This is important because it imposes the obligation to identify, measure, and then reduce to the minimum all risks from chemicals like pesticides in the workplace. This is an enormous advance in protection for workers, and it extends also to the public who may be exposed to the same risks.

Hazardous substances include any chemical that has the potential to cause any harm, from a cough to death. The risk is the chance of a substance causing harm in its normal use. This includes the toxicity of the substance, how it is used, how its use is controlled, who is exposed, etc. Pesticides are one of many substances covered by the Regulations.

Once any risks are assessed, employers must control them, either by removing or substituting the hazardous chemicals where possible, or by taking steps like isolating the chemical, installing ventilation, supplying protective clothing, or introducing safer procedures for handling the chemicals. Employers must then ensure that all controls are kept in efficient working order and monitor workers' exposure. Health surveillance is required where a possible exposure would be indicated by changes in workers' health. Symptoms like skin rashes and breathing problems could be monitored. Health records must be kept where monitoring is in force, and workers must be kept informed of the risks from their work as well as the results of the monitoring. They must also be told what precautions are taken on their behalf and about any they should take themselves.

The COSHH regulations take on board the requirement to wear protective clothing when using specified pesticides under the

Poisonous Substances in Agriculture Regulations, which were repealed when COSHH took effect.

Food Safety Act, 1990

This replaces the Food Act, 1984, and could be an important protector from the health effects of pesticides. It is an offence to sell food that is either 'unfit for human consumption' or 'injurious to health'. A food may be made injurious by 'adding any article or substance to the food [or] subjecting the food to any other process or treatment'. Pesticide residues should be covered by these criteria although they are not specifically identified in the legislation. The Act also says that, in deciding whether a food is injurious, due regard must be given not only to 'the *probable effect* [my italics] of that food' but also to 'the probable cumulative effect of food of substantially the same composition on the health of a person consuming it in ordinary quantities'. This would appear to include any residue for which there was any measurable risk of harm. Although not spelt out in the Act, MRLs and ADIs must be presumed measures of 'probable harm', but they are not measures of the smallest residues at which harm is 'probable'; rather those levels at which the degree of harm is no longer acceptable to the legislator. Thus *lower levels* of residue can also be shown to be 'probable' causes of harm and may be challenged under the Act. British law works by reference to the courts' interpretation of legislation and these points have yet to be tested there.

A more interesting section of the Act says it is an offence to sell any food that is not of the 'nature, substance or quality demanded by the purchaser'. These terms are not defined, but could be interpreted by reasonable people to include freedom from residues or at least some minimum levels. The Act also refers to the labelling of food in the same terms, making it an offence either to present for sale or to label any food in such a way as 'is likely to mislead as to the nature, substance or quality of the food'. Would this include selling or labelling a food without declaring any residues that might be present?

Although this section of the Act has not been tested in the

courts, it suggests that consumers can define their own quality standards as including zero pesticide residues and then encourage their Local Authority to take any supplier selling food with residues, or not so labelling the food, to court. Again, only a test case will resolve the question.

General laws and regulations

There is wide coverage of aspects of pesticide transport and use in general legislation. Acts such as the Consumer Protection Act, 1987, covers consumers against matters like misleading descriptions, defective products, etc. The Control of Pollution Act, 1974, covers depositing waste, and there is a Code of Practice covering disposal of empty pesticide containers and a Code of Good Agricultural Practice. Wildlife is protected in several Acts: the Wildlife and Countryside Act, 1981, lists protected species; the Animal (Cruel Poisons) Act, 1962, ensures that legitimate poisoning of animal pests is not unduly cruel; and the Animals (Scientific Procedures) Act, 1986, covers the use of animals in scientific tests for, among other things, pesticide safety.

REFERENCES

1. MAFF/HSE (1990). *Pesticides: code of practice for the safe use of pesticides on farms and holdings.* London: HMSO.

7 Food Residues

★ Pesticide residues cannot be removed entirely, but they can be minimised.

★ There is no such thing as a harmless level of residue, especially with carcinogens.

★ Although Maximum Residue Levels (MRLs) have been legally enforceable since 1988, and levels in excess are regularly reported by the government, there has not been a single prosecution.

★ Official methods of testing foods for residues are totally inadequate to detect or deter breaches of the law.

Food residues are a special pesticide problem. Spraying chemicals on farms or at home is voluntary and usually involves some degree of choice by the user about the safety precautions taken e.g. the specific chemical used, the protective clothing worn. Therefore any exposure to the chemical is also voluntary. Involuntary exposure during a country walk or while rummaging among recently treated timbers in the loft may be harmful and may not be avoidable, but they are still infrequent events. However, everyone eats food *daily* and consumes, without any choice in the matter, pesticide residues. Even organic food usually contains some pesticide residues. Food residues are the great leveller – whatever your occupation, income, overdraft, education, or eating habits, you are exposed to food residues every day of your life.

The best reason for being concerned about food residues is the closeness between humans and many of the targets of pesticides (see Chapter 1). This is brought firmly home, with a helping of humble pie, when we consider that many of our basic foods are exactly the same foods of important fungi, bacteria, and insects. Many otherwise beautiful insects are classified as pests because they attempt to eat our food before we can get our teeth into it. Indeed, pesticides are often used against them to give humans an unfair advantage in a simple competition between the pests and ourselves for food. Think of the slugs, aphids, and caterpillars hellbent on eating lettuce leaves before we can; of the weevils and fungi keen to devour bread wheat before we can. Think also of the cockroaches in our homes making a very good living off the very same diet as any British family. It puts a fresh interpretation on the saying 'you are what you eat'.

Residues are undetectable when food is bought and eaten – they have no smell or colour, and no taste. Considering the formidable biological action and wide range of toxicity among the chemical artillery thrown at pests, it is not surprising that public concern over food residues often provokes extreme reactions. Many consumers believe that *any* residue, however small, of *any* pesticide is an unacceptable risk to health. They want foods with the smallest detectable trace of a pesticide removed from sale. At the other extreme are the apologists for pesticides, usually government officials or chemical manufacturers, who assert that, so long as residues are below 'official' limits, foods 'will be completely safe for humans'. This view is often accompanied by a reluctance to question the validity of the safety limits or to spend adequate resources on monitoring residue levels in foods.

Both positions are unhelpful. Pesticides are used legally in the UK and other countries, and residues are inevitable. Some residues are impossible to eradicate. Even though DDT has been banned for a decade, all humans have its residues in their fat, and residues of both DDT and its breakdown products DDE and TDE are found widely in our foods. Any risk from residues depends on the specific chemicals, the amounts present in food, the amount of food consumed, and the susceptibility of the consumer. Rather than demanding a zero residue level, it is reasonable for consumers to demand that any residue levels tolerated in the food supply be both the lowest practicable and the least likely to cause any harm with any normal use of the foods containing them.

There are three essential steps to minimise food residues:

1   Ensure that only 'safe' chemicals are used in food production and that their methods of use minimise residues.

2   Establish maximum tolerated levels of residues in foods, based on strict safety criteria.

3   Maintain monitoring and enforcement procedures to ensure that excess residues do not enter the national food supply.

The rest of this chapter explores how British food supply stands up to examination under these criteria.

Minimising residues in food production

Consider how pesticides may leave residues on foods. Most residues follow on from the use of approved pesticides according to recommended or statutory conditions of application. Illegal uses and accidents do occur, but they merely add to the total load of residues. The persistence and nomadic properties of many pesticides used in agriculture is a sufficient cause of residues in harvested crops and livestock (see Chapter 2). However, there are also several aspects of pesticide application that contribute to residues while food is growing or while harvested products are moving along the food chain towards the dinner table.

Nomadic pesticides

As Chapter 3 describes, pesticides may move in the environment and end up in food. Water is an important route. Herbicides like atrazine leave negligible residues in crops, but seep into underground water or run off into rivers and can contribute significantly to the pesticide contamination of drinking water. Contaminated water used in food manufacture will transfer residues in the water to the foods. Another route is up the food chain into wildlife consumed by humans. Wood pigeons have DDT in their body fat. A MAFF study found samples that were higher than the EC Maximum Residue Level or MRL.[1] Similarly, high levels of lindane (HCH), dieldrin, and DDT have been recorded at several sites in freshwater eels. So high were these residues that the Working Party on Pesticide Residues was caused to comment that 'eels with high residues may present a potential threat to the long-term health of people who consume them in large quantities, to fish-eating birds and to wild mammals such as otters, which depend on eels as a major component of their diet.'[1]

Multiple spraying

The tighter regulations introduced since the Control of Pesticide Regulations, 1986, set the amount and frequency of pesticide applications to specific crops and give fixed withdrawal periods. The regulations are aimed at farmers using the minimum of chemicals to achieve the desired effect while being within some safety limits. Since much spraying of crops is undertaken on the 'better safe than sorry' principle, some farmers want to ensure that enough chemical is applied to *guarantee* no infestation. The regulations apply to each pesticide individually but can be overcome by farmers legally using several different pesticides of the same type at the same time. Such multiple spraying can be within the law but is contrary to its intentions and can increase the total amount of chemical applied to crops, thus increasing the chances of residues.

Shortening the gap between spraying and harvest

Most pesticides that break down take longer than 24 hours to do so. Therefore approvals to use specific chemicals on food crops often have 'withdrawal periods' stipulated in the conditions. This is a period of days or weeks before harvest during which no further applications of that pesticide are allowed. This aims to reduce the likelihood of residues being present by the time the crop is sold. Farmers have been known to ignore these withdrawal periods, spraying right up to harvest, especially with insecticides and fungicides, to minimise chances of blemishes that can slash the crop's value at market. Ignoring the legal withdrawal periods increases the likelihood of residues when food is eaten.

Spraying during storage and transport

Crops that are stored for long periods or that travel the world in ships from tropical farms to temperate dinner tables are often sprayed for protection in storage. Insecticides, fungicides, and rodenticides are the most common chemicals used. However, because such produce is often passed from trader to trader, and may be transferred from one storage site to another, the same consignment may be treated with pesticides several times. Although each application may be within the specified application rate for the pesticide, the residue level is increased when it is applied several times. Another source of residues is from pesticides applied to the walls and floors of silos and warehouses where grain is stored. Cereals, which are stored from harvest to harvest and are regularly moved from farm to merchants' storage, are therefore treated several times, especially with organophosphate pesticides like pirimiphos-methyl which are known to persist. One government committee has reported that 96% of grain stores use pesticides. The UK Agricultural Supply Trade Association (UKASTA) has recently introduced a voluntary record-keeping scheme where every batch of stored grain has its own record of pesticide treatments which travels with it when it is sold on or moved to different storage facilities.

Potatoes are also stored for long periods after harvest. To stop sprouting, tecnazene is applied to stored potatoes and has been found at levels higher than the EC MRLs. The amount was also greater than the originally proposed UK MRL that was dropped before the Regulations became law in late 1988.[2] (See page 122.) In August 1989, the MAFF reintroduced an MRL for tecnazene but set it at 5 mg/kg of potatoes, which is five times the original proposal of 1 mg/kg.

Illegal uses

Unapproved uses, either in choice of chemicals or in how they are applied, are an ever-present source of residues. Such illegal activity need not be common to be significant if it leads to doses of toxic pesticide being taken with food. Government monitoring of residues has found high levels of DDT in UK-grown cabbage long after DDT was banned in the UK. Such residues could occur only with illegal applications. A measure of the potential risk was given in a report on government pesticide monitoring covering the period 1979–80, which stated that 'The use of 43 individual pesticides has been recorded on glasshouse lettuce and one crop received 46 treatments involving 4 individual pesticides.' This crop takes only weeks from sowing to harvest, suggesting that pesticides could have been applied on a daily basis. Although regulations covering pesticide use have been tightened since this case, there is a very low rate of monitoring, which must fail either to detect or to deter such abuses, and it is not possible to say the extent to which it still occurs.

Pesticides on imported foods

Imported foods, both fresh and as manufactured products, can be a major source of residues. Processed pork and poultry from the People's Republic of China had high levels of DDT and lindane during several MAFF surveys in the 1980s. Levels regularly exceeded the EC MRLs. A study of pesticide use in Third World countries in 1988 showed that many countries use pesticides that are on a well-publicised UN list of pesticides banned in five or more countries. Many on that list are banned in the UK and yet have been shown to be used, often with few or no local controls or little or no adherence to recommended usages.[3] The implication is that high levels of residues of pesticides banned in the UK could be entering this country on imported fruit and vegetables or in other imported products like cocoa, tea, and dried pulses.

There is no systematic surveying of all possible pesticides on imported food. Only those pesticides *thought likely to be on food* are tested for in the UK, unless there is some scare about particular chemicals. Because many Third World countries have poor pesticide regulation and even worse monitoring facilities, it is impossible to know what pesticides might be used and hence what residues to look for. Many pesticide manufacturers in Europe and North America make and sell to the Third World pesticides that are either banned in their own countries or sometimes have never been offered for approval. So it is difficult to know what the chemicals are. Despite the fact that in the US it has been estimated by the Food and Drug Administration that 10% of imported food has illegal levels of pesticides, 70% of the 900 or so specified legal limits for cancer-causing pesticides in foods are never tested for.[4]

What is a safe level of residue?

There is no simple or reliable answer. There are internationally agreed upper limits for food residues, but they are not the same in all countries and not all pesticides have limits attached to them. Also, the way these limits are used often conveys a false sense of their being precise and scientifically sound.

Residues are measured by the ADI (Acceptable Daily Intake) and the MRL (Maximum Residue Level). The ADI is defined as the amount of residue, in mg/kg body weight, that, if eaten every day for life, should produce no harmful effects. It refers to the total diet and not to single foods. It is derived from the toxicity-testing data for pesticides. The 'no observable effect level' (NOEL) is taken as the starting figure. It is usually multiplied first by an arbitrary safety margin of ×10 because humans are said to be ten times as sensitive as the most sensitive animals, and then by another safety margin of ×10 because the most sensitive human is said to be ten times more sensitive than the average. This leaves an impressive, but still arbitrary, 'safety margin' of ×100.

Unfortunately, the NOEL level is only as good as the tests it is based on, and these have severe limitations when it comes to translating animal studies to humans. The tests are discussed in detail in Chapter 4. The whole notion of a NOEL for carcinogens and mutagens is ruled out by many scientists because, in theory, a single molecule of a carcinogen can cause cancer. So there can be no threshold dose *below which* there should be no effect. Therefore there may be no sense in setting an ADI for any pesticide that is even a suspected carcinogen. But there are more serious limitations to the ADI. It makes no allowance at all for possible interactive effects when more than one pesticide is eaten at a time. Any one food is likely to contain several residues and, of course, any diet is made up of many foods – perhaps 15 to 30 different foods are eaten in any week.

In addition, a safety factor of ×10 for human variation is only a guess. There is no way of knowing the real level of variation because the necessary studies would require large numbers of people and would not be permitted on ethical grounds. Even where there are some data on individual pesticides, there is no way of knowing the range of variation for all pesticides. Even were this known, no allowance is made for the needs of special groups like the elderly or children. And what of ill people whose sensitivity to all manner of environmental factors, from temperature to the contents of their diet, is increased?

The fundamental difficulty with the ADI and the NOEL lies in measuring the two types of data on which they are based – the amounts of residues in food and the effects on human health. Technology in chemical analysis has improved rapidly, allowing smaller and smaller amounts of residues to be analysed and identified with increasing accuracy. However, research on human health relies on techniques in toxicology and epidemiology which have not improved so fast. With human health, it is often difficult to know what to measure. It is also difficult to obtain reliable data for analysis. The major diet and health debates in industrial countries provide a good example. Despite the obvious importance of fat and sugar in the British diet, there are raging controversies over their role in many major diseases (e.g. heart disease, obesity, cancer), largely because of our inadequate understanding of disease processes and the difficulty of conducting precise experiments on humans. Most disagreements are about the quality of the data and not about the theories of diet and disease. The problems facing toxicologists working with food residues are immeasurably greater. They have to study the effects of minuscule quantities of chemicals, usually in the presence of many other chemicals in equally tiny quantities, the identity and behaviour of which may be unpredictable, on the health of humans over 10, 20, or 30 years, or even generations. Measuring *any* effect reliably presents major difficulties. In consequence there is always a great temptation to dismiss the health risk when tiny amounts of residue can be detected but there are no measurable ill effects. The wise err on the side of caution, giving human safety the benefit of doubt.

The MRLs are more vague and arbitrary in their formulation

than the ADIs, although both are based on calculated guesses. The MRLs are not meant to be used as safety limits, but are intended to be used as the maximum amounts of pesticide residue likely to be in specified foods produced according to 'good agricultural practice', which is defined as vaguely as the term suggests. Note that, unlike the ADIs, they are attached to individual food products and not the total diet. However, they take no practical account of how much of a contaminated food a consumer may eat in a day or month, or of the quantity of pesticides in the entire diet.

Despite the protestations of the MAFF and many in the food industry, MRLs must be measures of residue that are at least considered 'safe' or there is little point in having them. Indeed, the ADI is taken into account in setting a MRL, but exactly how is unclear because the criteria for setting each MRL are not published.

In setting an MRL, account is taken of any limitations attached to the pesticide's approval (application rates, withdrawal periods, approved crops, etc.) and of the general aspiration of farmers to use 'the minimum quantities necessary to achieve adequate control, applied in a manner so as to leave residues which are the smallest amount practicable and which are toxicologically acceptable'.

One of the great uncertainties in setting MRLs is the effect of any subsequent food processing on the residues or on their breakdown products. Very little is known about the effects of food processing on pesticides. Are any pesticides concentrated, diluted, or changed into less acceptable products by normal processing? UK brewers have found that the EBDC fungicides sometimes used on hops break down in the brewing process to less acceptable chemicals called thioureas. To protect their customers from a regular low dose in their pints, the brewers have imposed their own MRL of zero, even though there is no legal MRL for EBDCs on hops in the UK.

MRLs have been legally enforceable in the UK since 1988 but apply only to about 60 of the more than 400 active ingredients approved for use. The UK levels are based on the FAO Codex Alimentarius Commission (CAC) levels but are not identical. When the government circulated a draft of the Pesticides

(Maximum Residue Levels in Food) Regulations early in 1988, they published both a list of pesticides and a proposed MRL for each one. But by the time the Regulations had passed through Parliament, 7 pesticides had had their MRLs on certain crops, like lettuce and celery, mysteriously dropped from the list, while 11 had their MRLs raised to more easily achievable levels. Not one pesticide had its MRL lowered.

Pesticides and the crops for which there were proposed MRLs that were dropped before the draft passed into law are listed in the following table:

PESTICIDE	CROP
inorganic bromide	lettuce
dimethoate	lettuce
dithiocarbamates	celery, lettuce
iprodione	lettuce
haloxyfop	plums, strawberries, raspberries, blackcurrants, potatoes, beans, peas
tecnazene	potatoes
phosphamidon	oranges and other citrus fruits, apples, pears, peaches and nectarines, plums, grapes, strawberries, raspberries, blackcurrants, bananas, tomatoes, cucumbers, cabbages, Brussels sprouts, peas, lettuce

The adjustment of the originally published MRLs indicates the contribution of serendipity in the setting of MRLs. Why these seven were dropped after the normal consultations with 'interested parties' has not been disclosed. However, it is known that five of the dropped pesticides can exceed the proposed MRLs even when 'good agricultural practice' is followed strictly. They are tecnazene on potatoes, dithiocarbamates (EBDCs) on stem and leaf vegetables, dimethoate on lettuce, iprodione on lettuce, and inorganic bromide on leaf vegetables. How can we have any faith in the MRL system if established levels can be adjusted upwards at will, or, worse, dropped if they are difficult to achieve?

Since the 64 MRLs became legally enforceable in the UK, there has not (as of January 1991) been a single prosecution for a breach of this law. The explanation offered by a discreet official in the MAFF was that 'no court in the land would make a prosecution stick unless the residue was *three times the legal MRL* [my italics]'. The reason for this lassitude is that it is difficult to produce consistent results with the analytical techniques used to measure residues. A law designed to fix maximum residues but which can be enforced only when a residue is three times that maximum should be an acute embarrassment to the legislature.

Another difficulty with the MRL concept is deciding where in the food chain the MRL is to be determined. The UK legislation makes it an offence not only 'to leave' a residue in excess of a MRL but also to 'cause [a residue] to be left' in any food. Processing of farm produce may lead to concentration of residue in some foods. Residues that accumulate on citrus peel but not in the flesh will not be harmful if the peel is discarded. But if the peel is candied for cake making, the residue will be concentrated. Two MAFF scientists found that organophosphorus pesticides applied to wheat in storage can survive both milling and baking, so that up to 50% of the residue on the wheat remains in the bread if wholemeal flour is used. They concluded that wholemeal bread may exceed CAC MRLs and suggested that 'admixture of pesticides with grain could readily result in the present (CAC) MRLs [for bread] being exceeded. This in turn suggests that the maximum recommended doses for application to grain may be too high or that the MRLs are too low. However, there is little scope for reducing the application rates of many pesticides as this would seriously reduce their effectiveness . . . Some reappraisal of the MRLs for wholemeal flour and bread seems desirable.'[5] Is this a suggestion to raise an MRL that is difficult to stay below? This cavalier approach to MRLs came not from tendentious industry scientists but from scientists within the MAFF.

Despite the shortcomings of both MRLs and ADIs, such fixed criteria are essential for the successful regulation of residues in foods. However, residues below the MRL or ADI levels must not be presented as free from any health risks. There is always a risk with any level of residue, but the ADI and, by implication, the MRL are the arbitrary residue levels above which the risks are considered unacceptable. The more arbitrary the setting and application of these levels, the less confidence consumers will have in lower levels of residues being free from significant health risks.

Official doubts over MRLs and ADIs

The UK Committee on Toxicology of Chemicals in the Environment, which advises the MAFF on pesticide residues, showed clear caution with both ADIs and MRLs where children are concerned. Commenting on the presence of residues in infant foods as well as in eggs and milk, the Committee said, 'The residue levels in infant foods were generally low and unlikely to pose any hazard to health, but there are particular difficulties in interpreting data in terms of the health hazard to infants and young children and we consider that the grounds for prudence are particularly strong in this case. We therefore *recommend* that efforts be made to reduce residue levels in infant foods as much as possible.'[6] The same Committee expressed its reservations on the scientific basis for setting ADIs in commenting on the risks to infants and children of drinking milk contaminated with dieldrin, 'as the long-term effects, if any, of exceeding the ADI are unknown'.

The ACP, which issues approvals for pesticides, has also stated strong reservations about safety limits like the ADI or MRL. In explaining their decision to phase out dieldrin by the end of 1989, they said they 'acknowledged that the "drins" at any concentration in the environment were undesirable and considered that even the European Community limits did not provide total safeguards against the cumulative effects of the products'.[7]

The Committee on Toxicology has also expressed concern over the lack of recent studies of residues in breast milk. Breast milk is higher up the food chain than adult humans and could lead to breast-fed infants receiving concentrated residues from the mother's diet. Data on the effects of residues in breast milk on infants are scarce but should be studied more. A rapidly growing brain and spinal cord may be particularly susceptible to neurotoxic pesticides like the organochlorines that are known to accumulate in human fat tissue. A 1982 review of the Canadian and US scientific literature on DDT in breast milk showed that there had been no sign of a downward trend since 1970 when DDT was banned. Reported levels were so high they exceeded the WHO ADIs. A study of 102 samples of human milk in the UK during 1979/80 showed mean values of 0.007 mg/kg for HCH (lindane), 0.049 for DDT, and 0.002 for dieldrin. Mean values for HCH and DDT are greater than the current UK MRLs for the general population, and many of the samples of dieldrin also exceeded the UK MRL. It is important to initiate detailed studies to decide whether breast-feeding women need lower MRLs to ensure their milk does not carry excessive levels of residues.

Residues and the law

The regulation of food residues comes under two main bodies of legislation, the Food and Environment Protection Act (FEPA), 1985, and the Food Safety Act, which bundled the old food legislation, mainly the Food Act, 1984, and the Imported Food Regulations, 1984, into one body of legislation. Responsibility for enforcing food quality standards involving residues therefore lies with both the Ports Authorities and the Local Authorities, the latter through their Environmental Health Officers.

Under the Food Act, UK MRLs might be a criterion for food being 'unfit for human consumption' or 'dangerous to health'. It is not yet clear whether EC or CAC MRLs and ADIs will be accepted by UK courts for those pesticides (the majority) which are approved for use in the UK but which have no UK MRL. There is also a second form of offence, namely that of food being considered 'not of the nature, substance, or quality demanded by the purchaser'. Whether international MRLs would be accepted here is also uncertain. It is interesting to note that a Local Authority has to decide whether a food is 'unfit' or 'injurious', but that consumers can decide whether a food meets acceptable criteria of 'nature, substance or quality'.

Under Part III of the FEPA, the Pesticide (Maximum Residues in Food) Regulations, 1988, it is an offence to cause or to have left in specified residues exceeding the UK MRLs laid down for 64 pesticides. According to the wording of this legislation, it is sufficient to exceed an MRL for an offence to be committed.

Monitoring the food supply for pesticides

The only way to discover whether there are unacceptable levels of residues in foods is to conduct systematic monitoring of the food supply. This is no minor undertaking. Not only is there a large number of different foods consumed in the UK, but there are also thousands of farmers, distributors, manufacturers and processors, caterers, and retailers.

There is no systematic residue-monitoring programme in the UK. The little official monitoring that does occur is inadequate to assess whether pesticides are being used according to the law, and this does not encourage confidence in the safety of the British food supply.

There are two sources of official residue testing in the UK:

The Ports Authorities and the Local Authorities

These agencies enforce the food safety legislation. Both are generally reactive, checking suspected shipments of food or acting in cases where there is suspicion of residues. Samples are analysed by the local Public Analyst, whose results are accepted in courts of law. The number of samples taken for residue analysis is very small. One national estimate from the Institute of Environmental Health Officers put the number of samples analysed for pesticides in 1982 at 121 out of a total of 20,000 analyses undertaken for all forms of food contamination. Ports Authorities are the main opportunity to check imported foods for residues. Limitations in resources are a significant constraint on both the number of samples taken and the number of pesticides tested for. The House of Commons Agriculture Committee was told in evidence that some ports 'never take any residue analyses while in others only a narrow range [is] ever tested, some 20 of the more than 400 active pesticide ingredients approved for use in the UK'.[8] The Association of Port Health Officers expressed their concern and gave an understated warning that 'the fact that a pesticide is banned in this country does not mean we are not getting residues on our food'.

Local Authorities face similar constraints. Most Environmental Health Departments are understaffed because of a shortage of resources. Pruned budgets also mean that samples can be sent to the Public Analyst only when the officers are so sure of a conviction that they can justify the cost of collecting data because it is needed for court evidence. One survey showed that in 1983 the County of Avon analysed fewer than 200 samples, and the County of Kent only 50. Resource shortages reduce the number of samples tested and the number of pesticides looked for during analysis. In short, Local Authorities do not have the resources to carry out monitoring of residues in food. Their reactive sampling must not lead to the conclusion that there is little problem with excessive pesticide residues. They have not carried out the necessary analyses on which to base such confidence.

A measure of the contamination found through the Public Analyst's work is shown in two special studies of residues in fruits and vegetables around 11 counties of England in 1983. The first survey involved specially collected random samples of fruit and vegetables, not samples that were suspected of being contaminated. Of the 23 different fruits sampled, 14 were found to have some residues. Nearly half of the samples of pears, tomatoes, strawberries and raspberries were contaminated, and the number of pesticides reported for any one sample varied from 1 to 7. Of the 34 types of vegetable sampled, 18 were contaminated with up to 8 different pesticides. Taking all samples of fruit and vegetables, one-third had detectable residues.

The second survey was a collaboration with the MAFF to look at residues of DDT and its isomers, DDE and TDE. Residues of DDT, or its isomers, were found in 10% of samples tested, including products for which the use of DDT was no longer approved.

The Working Party on Pesticides Residues (WPPR)

The second source of residue testing is that carried out on behalf of the MAFF's Steering Group on Food Surveillance. The testing is conducted through a Working Party on Pesticides Residues

(WPPR) that was set up in 1977. The original purpose was to
monitor pesticides covered by the old voluntary Pesticides Safety
Precaution Scheme. This meant looking only for the limited list of
pesticides covered, and the WPPR declared pesticides on imported food to be outside its terms of reference. In fact, as recently
as 1982, its task of monitoring for residues was seen to be
relatively unimportant compared to regulating the use of
pesticides on farms (see page 109). In their annual report 1977–
81, the WPPR said: 'Problems arising from the presence of
pesticide residues in food are fortunately rare, but when they
arise this is generally not as a result of negligence but rather
because of the unforeseen effects of normal use. In these circum-
stances it is more appropriate to give advice or direction which
can be tailored to the specific problem, rather than resort to a
court of law.'[9] The same report aired the working party's disap-
proval of MRLs, which the EC was then leaning on the UK to
introduce. The report said: 'A statutory scheme, based on the
enforcement of MRLs, could result in food being condemned as
unfit if it contained a higher residue than the MRL even though
overall exposure to the residue might be very low. In most cir-
cumstances, occasional exposure to higher-than-average levels of
a pesticide in a foodstuff has no public health significance'. When
the monitoring body has that view of MRLs, their diligence in
searching for excesses must be questioned.

The WPPR does not monitor the complete national diet at a
sufficiently intense level of sampling to give any useful estimate
of the level of residues across the diet. It concentrates on both a
'rolling cycle' of surveys of the major food commodities and spot
surveys whenever special problems arise. It has, for example,
surveyed DDT in fruit and vegetables, bromide in food crops, and
post-harvest fungicides in potatoes. Recently it has undertaken
continuous monitoring of selected foods, starting with bread,
milk, and potatoes in 1987, and adding other cereals, animal
products, and fruit and vegetables in 1988.

An important part of the regular sampling of foods for analysis
is the UK Total Diet Study, which began in 1966. Despite the hint
in the title, this is not a residue survey of the Total Diet, but a
sample of the national diet from which selected residue studies
can be made if and when convenient. The sample has changed

slightly over the years. It used to be divided into seven groups of foods, the quantities of each bought being based on the national average diet as measured in the National Food Survey conducted by the MAFF. In 1975 this was increased to nine food groups, but since 1981 there have been 20 food groups available for analysis – bread, miscellaneous cereals, carcase meat, offals, meat products, poultry, fish, oils and fats, eggs, sugars and preserves, green vegetables, potatoes, other vegetables, canned vegetables, fresh fruit, fruit products, beverages, milk, dairy products, and nuts.

Since 1981, a total of 115 different foods has been bought at one of 24 different locations around the UK once every 2 weeks, 24 times a year. This gives 24 'baskets' of food each year. The amounts of food bought are about what an average adult would consume in a week, based on the National Food Survey data. The 24 samples of food are taken to a laboratory in Norwich where they are prepared and cooked as they would be in the home, and then collected into their 20 groups and mechanically homogenised in a blender before being frozen in case they are wanted for analysis.

The sample may be a statistically interesting sample of the nutritional composition of the UK diet, but it has shortcomings as the main basis for studying pesticide residues. Take green vegetables and, in particular, cabbage. Only about 5 oz (144 g) of cabbage is collected at each of the 24 samplings throughout the year. At best, this means that only 24 different cabbages are sampled all year to check residues in the whole country's supply of cabbage. All cabbages are reasonably similar in their content of potassium, protein, or vitamin C, which are natural components of cabbage. But they will vary dramatically in their residue content, which depends on the grower. Even if a badly contaminated cabbage were to be chosen, its blending with all other green vegetables sampled *before* analysis would dilute the concentration of the residue, making it appear less significant in the analysis of the whole sample.

The overall effect of the Total Diet method is to dilute the evidence of residues. Most pesticide residues are very unevenly distributed through foods. If a pesticide is used by almost all growers on a particular crop, and they all use it according to the instructions, the residue level might be expected to be similar in all samples of the crop. However, growers use different pesticides, they do not all follow the approved procedures, and the UK diet includes a large volume of imported foods containing unknown residues. So any useful monitoring must be able to detect and measure these variations.

The Total Diet also makes unacceptable assumptions about the different eating patterns in the UK. Not only do the Scots eat a diet different from that of the Welsh or of the English, but within any group – national, ethnic, age, income, etc. – there are dramatic variations in eating habits. This is where pesticide residues really begin to matter. Children drink more milk than average, people on a low income eat more bread, young adolescents may eat more potato products (crisps, chips), and vegetarians and vegans eat much more fruit and vegetables and more imported pulses and nuts. The study can make no allowance for these variations. Nor can the analysis of samples be adjusted easily to accommodate them.

Take wholemeal bread, which has been shown to harbour organophosphate residue in the bran. It occupied only 15% of the weight of the bread group sample in 1987 but is known to be the source of most bread residues. Eighty-five per cent was white bread, which is known to have lower total residues. So the total residue level indicated by the survey hid the fact that those people following health advice to eat wholemeal bread rather than white had their intake of residues underestimated. In addition, the sample of bread in the Total Diet reflects only the national average bread consumption. Some people eat much more than others.

Lettuce are another known source of residues which can exceed the CAC MRLs under approved conditions of use. Regular salad eating could increase the residues of the insecticide dimethoate and the dithiocarbamate fungicides eaten. However, because lettuce are homogenised with all other leaf vegetables before analysis, the specific risk from lettuce would not appear in an analysis of the Total Diet samples.

In its 1989 report, the WPPR reported on a study of organochlorines in the nuts group. Residues were found, but 'the overall estimated dietary intake of organochlorine compounds was not greatly affected due to the small contribution this food makes to the average diet'. That statement was based on the average UK diet which contained only 0.57 oz (16 g) of all nuts and nut products per week. This is equivalent to less than a small packet of peanuts. Many people with an ordinary diet eat more than this, and vegetarians eat ten times as much, and more. Anyone eating more than the average could be at much greater risk but is neither warned, nor advised about which of the many available types of nut are likely to have high residues.

The UK Total Diet Study *could* be a useful measure of *trends* in residue levels in its specific, limited sample of the average UK diet. For this to occur, the whole of the sample would need to be analysed annually for all possible residues, or at least for the same set of residues each year. At present this does not happen. Only some food group samples are analysed each year, and then for specific pesticides. However, the study is being used inappropriately to assess residue *risk* in the food supply. This inevitably plays down the amount and importance of any residues present. The WPPR's own studies on specific foods regularly indicate dangerous residues which are being masked in their studies of the Total Diet samples. It is a dangerous proxy for a full monitoring system which could aim at the individual foods in the national diet and provide the means of assessing risk among the range of different diets in the population. Full monitoring of individual foods would also allow the MRL legislation to be implemented more effectively and give practical data to help guide future pesticide research and regulation.

The WPPR's approach to monitoring

Analyses for residues are carried out either on the Total Diet samples or on specially collected samples in special studies. The results give some interesting insights into the levels of residues but should be interpreted in the light of several factors:

Number of samples analysed

The total number of samples analysed is small. Baroness Trumpington, Agriculture Minister in the Lords, told Parents for Safe Food in 1989 that 'monitoring of residues by the Working Party on Pesticide Residues clearly cannot cover every product but some 2–3,000 samples are analysed each year for a wide range of pesticides'.[10] The 1990 report said that 4,000 samples were analysed but these were conducted over the years 1988 and 1989.[11] If this number covers all food samples throughout a whole year, it represents an average of fewer than one sample of each main food product each year. In fact, not all products are sampled. The Working Party has limited resources and tends to 'focus resources on those areas where residues are most likely to be present'.[12] This is more case finding than monitoring, and continues despite the Working Party's being continually surprised by the unexpected levels of pesticides discovered in its surveys.

Number of pesticides looked for

Analyses do not look for all UK approved pesticides. Many studies look only for one group of pesticides, for example, organochlorines or organophosphates. Other studies identify multiple residues, but the maximum number of pesticides covered is only between 40 and 70, depending on the combinations of testing techniques. This is only a small proportion of the more than 400 pesticides approved in the UK, in addition to others, which, though not allowed here, could be arriving on dinner tables via imported foods.

Low sensitivity of tests hides residues

If no residue is detected, it doesn't mean that none is present. There is a limit to the sensitivity of the available analytical techniques, and for some pesticides there is no suitable analytical technique at all. The lowest concentration that can be detected is called the limit of detection. This varies with the technique and thus with the laboratories doing the testing. Often, several laboratories are used to check test results. To allow for these differences, the MAFF use a 'reporting limit' which is the smallest amount of pesticide detected by the *least sensitive* method in any of the laboratories used. Any residues found below this limit are recorded as ND or 'not detected' in their results, even though they may be both present and measurable in some laboratories used by the MAFF.

Not all residues are harmful

The presence of residues does not mean the food is harmful. There is great uncertainty over how a residue is to be judged 'safe'. The WPPR follows UK MRLs where they exist and CAC or EC MRLs where there are no UK equivalents. Interpretation of their residue analysis relies on the availability of toxicological information, and the Working Party is open about the limitations of giving a clean bill of health to most of its results since, 'at present toxicological interpretation is particularly difficult where residues of more than one pesticide are found in the sample.'[13] Most of its analyses show more than one residue present.

Selected WPPR residue survey results

Organochlorine pesticides have been a constant cause of concern to the WPPR since 1966. The 1979–80 analysis of 22 samples of the Total Diet showed most residues to be of lindane, dieldrin, and the DDT isomer, DDE. Since the banning of DDT, dieldrin, hexachlorobenzene, and lindane on food crops, residue levels have fallen. However, the analysts found high levels of dieldrin in some milk and potato samples; of lindane in some meat; and of pentachlorophenol in samples of meat, milk and poultry. While the average level in a group of samples may be low, the fact that there are a number of very high levels suggests that residues are highly variable among samples.

Some of these levels exceed the MRLs. Dieldrin at the level of the MRL was found in some of the 75 samples of main-crop potatoes collected in 1987. Samples of strawberries, Brussels sprouts and cabbage taken in the years 1985–87 had DDT levels in excess of the MRL. These are worrying results because they suggest that farmers have been using DDT illegally. Liquid milk analyses also found levels of lindane above the MRL and levels of dieldrin close to the MRL, and in one sample of UK cheese and one of Scottish cheese, the MRL for lindane was exceeded. The Committee found these dairy product residues 'unexpected both in terms of the levels found and their incidences'. They also indicate a possible hidden problem of pesticide abuse, because analysing a small number of samples gives no representative picture of all dairy farmers. This may lead to large under- or overestimation of residues in the food supply. Only 20 samples of Scottish cheese (of which 5% were above the MRL for lindane) and 105 of cabbage (of which 4% were above the MRL for DDT) were tested in typical studies of the national supply. Much larger samples may be necessary to discover whether these few excessive levels are rare coincidences or exist throughout the national production. In 1990, the WPPR took only 126 milk samples and found 12% had organophosphate residues. Only 39 samples of UK-grown cabbage were analysed, but no studies on Scottish cheese were reported.[11]

Regular surveys of potatoes were begun in 1987 because they are such an important staple food. However, a special study in 1985–86 that was looking at post-harvest treatments on a range of fruit and vegetables found widespread residues and serious breaches of MRLs on potatoes. The crops analysed were, in addition to potatoes, apples, pears, and white cabbage. Eight different pesticides were found. Fifty per cent of the apples, 33% of the pears, over 76% of the potatoes, and 73% of the cabbage were found to contain one or more residues. The study found that 37 out of 67 samples of UK potatoes had tecnazene residues above the CAC MRL. This was one of the first reports of studies showing that even when tecnazene was applied according to 'good agricultural practice', it still left excessive residues. In 1987, three samples exceeded or equalled the MRL. The ACP commented that exposure of children to levels of tecnazene at the CAC MRL (1 mg/kg of potatoes) 'could lead to exposure 13 times the ADI' but felt there was too high a safety limit included in the ADI. In 1989, the MAFF introduced a higher UK MRL of 5 mg/kg for tecnazene.

The 1990 report showed a great improvement in residues of tecnazene in the 126 samples of UK potatoes but failed to make clear that the goal posts had been shifted. The MRL had been increased from 1 to 5 mg/kg since the previous report.

Peeling potatoes can remove up to 90% of the tecnazene, but cooking removes little more. The hazard of excess residues is therefore the greater for all those healthy eaters who have increased their consumption of jacket potatoes or who have taken to boiling potatoes without peeling them.

Infants and young children eat greater amounts of food for their body weight than adults and are thus more susceptible to residues. A special analysis of seven types of infant foods was made in 1985 and 1987 because there were hitherto few data on residues in their typical foods. Organochlorines and organophosphorus pesticide residues were looked for in infant milk formulas, rusks, and prepared baby foods based on cereals, meat, fish, or eggs, as well as fruit and vegetables. Two of the milk formulas contained residues of lindane. The rusks and other cereal-based infant foods contained the same range of organophosphorus residues found in general cereal products like bread. Several samples had more than one organophosphorus residue (pirimiphos-methyl, etrimfos, chlorpyrifos-methyl). The levels of pirimiphos-methyl were higher than the current UK MRLs.

Conclusions

The results of the small amount of residue testing conducted for government departments do little to support the Minister of Agriculture's belief that 'our food contains very low levels of pesticide residues'.[14] While much of the food sampled does have low levels of residue, the results also show significant numbers of samples with excessive residue levels as measured by existing MRLs. Noting that not all foods are analysed and a very small proportion of possible pesticides are analysed for, it would seem appropriate for the MAFF to conduct a full review of the residue-monitoring arrangements.

Not only should the sampling rate and the number of pesticides analysed for be increased, but the number of pesticides with UK MRLs must be enlarged to bring the UK in line with the EC lists of MRLs. At the very least, *every* approved pesticide should have an MRL. The relationship between the ADI, or at least the safety component used, and the MRL should be made clear. MRLs should be enforced at the level set in the same way that blood alcohol levels are in drink-driving offences. There is no scientific basis for 80 mg/100 ml blood being the drink-driving limit and, in fact, there are many who argue that it should be lowered to 50 or 60 mg/100 ml to reduce road-traffic accidents. The legal limit is just what it says and is applied accordingly. Register 80 mg or more on being breathalysed and an appointment with the magistrate follows. It should be realised that a level of 75 or 50 mg does not mean that a driver is sober enough to be a safe driver. Similarly, a residue just below an MRL should never be assumed to be safe.

It also seems against both the principle of an MRL's being rooted in 'good agricultural practice' and the MAFF's professed interest in safe food for MRLs ever to be increased. As pesticides become safer in themselves and as farming methods improve, less and less pesticide should be required. In response, the MRLs *should be lowered, and lowered progressively*, to reflect continuously the smaller residues achievable by good and efficient farmers. The pesticide manufacturers argue that the latest products are more specific, require less chemical to be sprayed, and break down faster and more safely in the environment. There are also signs that existing pesticides can be used at much lower rates of application without any loss of yields. Research in Scotland has shown that cutting herbicide applications by 87% has not reduced yields of wheat and barley.[15]

The Minister has taken steps to have data from the major food retailers' own independent residue testing (see Chapter 8) made available to the MAFF. While it is useful to share such data, there is no alternative to the government's ensuring the adequacy of its own monitoring. There is a potential danger in relying too much on the retailers' data: they conduct more residue analyses each year than does the public sector and, because of their tough contracts with both growers and suppliers, the retailers' food residue data appear to be well below the national average. The more they are blended into the national data on residues, the more they will lower the national average and convey a false picture of the general level of residues.

The Minister of Agriculture has often said that 'There can be no compromise on food safety. The Government must put safe food first . . . The best way to see that the British people choose British food is to uphold the highest possible standards. We must be in league with the customer if we are to serve industry well.'[16] Improved residue testing, more publicity for the results and specific action to lower the level of residues in food would be considered by most people as solid steps towards the 'highest possible standards'.

REFERENCES

1. MAFF (1989). Report of the working party on pesticide residues (1985–8). Twenty-fifth report of the Steering Group on Food Surveillance. *Food surveillance paper no. 25.* London: HMSO.
2. MAFF (1989). Op. cit.
3. Pesticides Trust (1989). *The FAO code: missing ingredients.* London: Pesticides Trust.
4. Weir, D. and Schapiro, M. (1981). *Circle of poison.* San Francisco: Institute for Food and Development Policy.
5. Wilkin, D. R. and Fishwick, F. B. (1981). Residues of organophosphorus pesticides in wholemeal flour and bread produced from treated wheat. *Proceedings of the British Crop Protection Society conference. Pests and diseases,* pp. 183–7.
6. MAFF (1989). Op. cit., p. 61.
7. ACP (1990). *Advisory committee on pesticides annual report 1988.* London: HMSO.
8. House of Commons (1987). *The effects of pesticides on human health.* Vol. 1. Report and Proceedings of the Agriculture Committee. Second Special Report, Session 1986–7, para. 147. London: HMSO.
9. MAFF (1982). Report of the working party on pesticide residues (1977–81). Ninth report of the Steering Group on Food Surveillance. *Food surveillance paper no. 9.* London: HMSO.
10. Open letter to Parents for Safe Food, 7 June 1989.
11. WPPR (1990). *Report of the working party on pesticide residues: 1988–9.* London: HMSO.
12. MAFF (1989). Op. cit., para. 10.
13. MAFF (1989). Op. cit., para. 7.
14. MAFF, press release, 3 April 1990.
15. *Farmers Weekly.* 6 April 1990, p. 14.
16. MAFF, press release, 27 September 1989.

8 Food Industry Response

- ★ Farmers have begun listening to consumers and environmentalists, and are trying to reduce their dependence on pesticides.
- ★ Supermarket monitoring of pesticides is more intensive than the government programme.
- ★ Supermarkets have stricter controls on pesticide abuse and enforce lower levels of residue than the government.
- ★ Foods rejected by supermarkets for high residue levels may enter the high-street and barrow trade.
- ★ Government efforts to control and reduce pesticide use should match and exceed the higher standards already found in some sectors of the food industry.

The relationship between the food industry, particularly the retailers, and consumers has never been so close. To say the liaison is cosy would be an exaggeration, but consumers are being listened to as never before, and this opens great opportunities to influence the type and quality of food placed on supermarket shelves. The explanation lies in the changing structure of the food industry.

Put simply, there have been three distinct revolutions in the food industry since 1950. The first is the reconstruction of farming, with increasing modernisation and attempts to grow more food in the UK. In the early 1960s UK farmers, who had strong support from the government, and the world agricultural trade determined what food was on the UK table. At that point the food-manufacturing industry was in its infancy, but it expanded rapidly and led to the second revolution. More and more house-hold food purchases were of processed or manufactured prod-ucts. Fresh farm food was seen less in shopping baskets and was being replaced by tinned and bottled food as well as pre-mixed meals and, more recently, ready-to-eat meals and frozen foods. The power of the farmers' lobby began to give way to the buying power of the large manufacturers, who used farm produce more as a manufacturing resource than as a food. Consumers bought their food from corner shops which had no control at all over the quality or types of food offered by the manufacturers. Throughout these two revolutions shoppers had little influence over their food supply.

The third revolution developed rapidly from 1970 and now dominates most decisions in the food system. It was the growth of supermarkets and especially the growth of a small number of giant chains or multiples. The buying power of retailers began to dominate the manufacturers, the farmers having already become very weak bargainers.

Today, the large supermarkets dictate to the rest of the food industry. They decide what goes on their shelves and even influence the composition and packaging of the food they sell. They sell branded foods but apply additional pressure on brands when they sell alongside their competing own-label products. An interesting twist is the return to more fresh foods. The develop-ment of supermarkets has ensured that these are available year round, in more variety and freshness than ever before. Con-sumers now buy direct from the most powerful link in the food chain and their influence on the supermarkets comes from the rivalry and competition between the chains.

Supermarkets are successful because of their high turnover. But high-volume sales require large numbers of shoppers going through the checkouts. Because there is competition among the top five – the Co-Op, Safeway, Sainsbury's, ASDA, and Tesco – and their margins are below double figures, they need each and every shopper they can coax into their stores. They also want customers to return each week, and their public-relations efforts concentrate on building customer loyalty. This competition for customers is the source of the new consumer influence, which can be described as the closest the shopper has come to controlling the food supply since the Industrial Revolution.

There are clear examples of the supermarkets responding to consumer wishes. When the campaigns against food additives raged in the 1980s, supermarkets were quick to act. First they began dropping unnecessary additives from their own-label products, and then proceeded to influence the manufacturers of branded foods. Safeway announced the removal of 23 'cosmetic' colouring and flavouring additives in one blow. Nutritional labelling of foods was pioneered by supermarkets in the early 1980s. Consumer campaigns were increasing in strength but had still not managed to get the government to require manufacturers to include nutritional information on food labels in addition to the ingredients lists which were already required by law. The supermarkets saw the customer interest as a marketing opportunity. In 1983, Tesco launched a detailed voluntary labelling scheme on own-brand products and, within two weeks, Sainsbury's followed. Now all major supermarkets and most large manufacturers have some form of nutritional labelling on products. When food irradiation developed into a consumer issue in 1987, several supermarkets responded quickly by stating that they would not sell irradiated food even were the government to legalise it. There is little doubt that this cautious stance dulled the enthusiasm with which the MAFF and some other sectors of the food industry were rushing to have irradiation legalised in Britain.

A cursory analysis of supermarket action on these issues shows that the retailers were not only more rapid than both the manufacturers and the government in their response to public concern, but were also more prepared to make the necessary changes. Pesticide residues in food are a central concern of consumers, and the food industry is already beginning to respond. Again the government is being shown slow to respond to demands for higher standards, and the retailers and their suppliers appear to be taking the initiative.

The food industry's action on pesticides is motivated by three central factors. First, the new pesticides legislation passed since 1985, especially that on maximum residue levels in foods: the existence of MRLs means that residues can be judged against an accepted benchmark.

Second, the new Food Safety Act, 1990 (see page 111), introduces the useful concept of 'due diligence'. In essence, this means that a seller of food can no longer blame a dangerous level of residues on the person who supplied them, be it a wholesaler or a farmer. The seller will have to show that they took all reasonable steps to ensure that the food sold was not unfit, contaminated, or injurious to health. This means the supermarkets will have to check that their suppliers are complying with the law, and could be expected to conduct their own residue testing as well. The law doesn't define 'due diligence', preferring to leave it ominously vague, like the Scouts motto, 'Be prepared'.

Third, and most feared, is the aware and concerned consumer. Unlike the agrochemicals industry and many government departments, the food retailers listen to and respect consumers, even though they may not have the best information, or be 'experts' on pesticides.

There are signs that the industry's response to consumer interest in pesticides will be more thorough than on any earlier issue. This time much effort is going into trying to anticipate and act on concerns before they become coordinated public campaigns. History seems to have taught harsh lessons. The speed and impact of consumer resistance to food additives caused much chaos in the food industry. The industry was not used to being at the receiving end of market demand and was ill-prepared for what followed. Long-established consumer purchasing patterns were changed within months, and both manufacturers and retailers were left with shelves full of unwanted, additive-rich foods and a shortage of the preferred alternatives. Products had to be reformulated quickly, new labelling designed, and marketing and advertising plans thrown back to the drawing board.

Consumer reaction on additives was swift and had an early impact on the market. There may have been a consumer over-reaction, with many shoppers shunning all additives and rejecting products with the most E-numbers in their ingredients list, irrespective of what they were. In retrospect, however, it seems that any overreaction was fuelled by secrecy. Neither the MAFF, which regulated approved additives, nor the industry was prepared to release information on either safety testing of additives or the criteria for allowing individual additives. Secrecy fuelled suspicion, and, without any information on which to judge the acceptability of additives, it was logical for consumers to reject these unnecessary and unknown risks to health. Overreaction was inevitable, given the lack of cooperation from government and industry.

There are many similarities between this issue and the secrecy surrounding pesticides. With pesticides, the industry still has the opportunity to share information with its customers and avoid similar overreaction and chaos. A special survey was conducted for this book of major retailers and leading food industry organisations in order to discover their policies on pesticides and to see what actions were being taken to protect consumers.

The *Poisoned Harvest* survey on pesticide policy and action in industry

Much has been written about what the food industry does regarding the use of pesticides on the food supply. This is often critical, and sometimes justly so, but insufficient account seems to have been taken of what the industry itself thinks and says. The survey was planned to discover the thinking and actions of leading companies or organisations in the food chain. The closer the food industry is examined, the more interesting and constructive its role in controlling pesticide abuse appears.

The survey was based on a detailed questionnaire which was followed up where appropriate by interviews. The survey sample was designed to reach a broad range of the major food-retailer groups across the country plus a number of leading trade organisations. The retailers sample includes the 'Big Five' – the **Co-Op**, **Safeway**, **Sainsbury**, **Tesco**, and **ASDA** – and Waitrose, together with a range of the smaller retailers, including those operating chains of small grocer shops, across the UK. The supply or manufacturing side was represented by a sample of the leading trade or producer organisations, covering both the main food production sectors and manufacturers.

The companies or organisations to which questionnaires were

sent are listed below. The first mailing was in November 1989, with a reminder sent out in December. Those in bold type returned either completed questionnaires or letters explaining their position on pesticides. The mailing list was compiled as much to identify who took consumer concerns seriously as to discover what actions were being taken. The relatively small number of replies says much about the food industry's interest in its customers. The responses from the dominant retailers and organisations like the **National Farmers' Union (NFU)**, the **Food and Drink Federation (FDF)**, and the **Brewers' Society** include the leaders in key grower and manufacturing sectors of the food industry.

Food retailers

Argyll Foods plc
ASDA Group plc
International Stores Ltd
Bejam Freezer Food Centres Ltd
Budgen Ltd
British Home Stores plc
Co-operative Wholesale Society Ltd (CWS)
Cullens Stores plc (NISA)
Dee Corporation plc
Frank Dee Supermarkets Ltd
Fine Fare (Holdings) Ltd
Hanburys Ltd
Haslett, J. & J. Ltd
Hillards Supermarkets Ltd

Holland & Barrett
Kwik Save Discount Group Ltd
Littlewoods Organisation plc
Londis (Holdings) Ltd
Low, W. M. & Co. plc
Mace Wavy Line
Marks & Spencer plc
Morrison, W. M., Supermarkets plc
Safeway plc
J Sainsbury plc
Spar (UK) Ltd
Tesco plc
Waitrose Ltd
Woolworth, F. W., plc

Industry organisations

Association of Cheese Processors
Association of Independent Retailers
BACFID
Bacon & Meat Manufacturers' Association
Biscuit, Cake, Chocolate & Confectionery Alliance
Brewers' Society
British Farm Produce Council
British Frozen Food Federation
British Independent Grocers' Association

British Poultry Federation
British Retailers' Association
British Turkey Federation Ltd
DAFFA
Dairy Trade Federation
Edward Vinson Limited
English Country Cheese Council
Flour Advisory Bureau
Food and Drink Federation
Food From Britain
Fresh Fruit and Vegetable Information Bureau

Health Food Manufacturers'
  Association
Humber Growers Limited
London Herb and Spice Co. Ltd
**National Association of British &
  Irish Millers Limited**

**National Association of Master
  Bakers, Confectioners & Caterers**
National Dried Fruit Association
National Farmers' Union
Potato Marketing Board
Tea Council Ltd

Policies on pesticides

A clear difference in approach was evident between the industry
organisations and the retailers. The former generally took current
legislation as their policy, which may reflect their close association with the MAFF and other government departments during
the drafting of legislation. The retailers were much more progressive but, with rare exceptions, were neither growers nor manufacturers of foods and could take the consumer's interests on board
more easily.

Among the industrial organisations, the **NFU** has the clearest
and most predictable policy on pesticides. Chemicals were seen
to be necessary for farmers because they 'help increase yields,
improve productivity, and maintain farming incomes'. The **NFU**
follows existing legislation but also supports 'moves to optimise
and ultimately reduce the use of pesticides [and] efforts to find
more environmentally friendly pesticides, reducing dependence
upon those which produce harmful residues'. Pesticides are seen
as an important part of farming, but farmers are more exposed
than most of the population, and it is a healthy sign that the **NFU**
joined with the **BAA** (British Agrochemicals Association) and
environmental groups in 1989 to press the government to speed
up the testing of new chemicals (see page 88). The **NFU** is less
progressive, however, when it comes to consumers. Pesticides
are said by the **NFU** 'to provide the consumer with the quality and
quantity of foodstuffs they demand. Modern consumers do not
want blemished or pest-ridden produce.' There was no mention
in the statement of health risks from residues and no recognition
of the growing pressure for suspect chemicals to be dropped
where they are not essential.

The **FDF**, which represents most of the large manufacturers in
the UK, also took current legislation as its frontline position but
accepted that individual members could do what they pleased.

The **FDF**'s interest in pesticides policy is recent and 'has developed from the time that FEPA was under discussion', which was around 1983.

Organisations like the brewers and flour millers often have specific interests in pesticides because they are used in their processing methods. The **Brewers' Society** developed an active interest in pesticides as early as 1980. Because brewing uses live yeast during fermentation, there was understandable concern over the possibility that any fungicide residue might disrupt beer making. In addition, the chemicals sprayed on both the hops and the barley are a source of flavour taints. However, in 1980 the Society also expressed concern for the health of beer drinkers, a significant proportion of the population, who take a regular part of their calories from brewery products. They accepted that 'a large section of the population may thus be exposed to sub-acute but regular doses of residues or their uncharacterised degradation products'. As a result, the Society began warning the MAFF and the **BAA** of any pesticides that disrupted their product or were suspected to be unsafe and reserved the right to 'refuse to buy barley or hops known to have been treated with the [suspect] chemical'. In addition, members were advised to request certificates from all suppliers guaranteeing that neither hops nor barley were treated with any nominated suspect chemicals.

By 1990, the Society had strengthened its concern with pesticides by publishing, for hop growers, a list of the MAFF approved pesticides for use on barley and hops, indicating those that had been tested and approved by the Society. This, in effect, recommended growers to use only certain of the legally allowed chemicals. This initiative has spread and is controversial. There are companies who feel that any grower should be allowed to use any legally approved pesticide, while others assert their right to choose what chemicals from the approved list they will have on their products.

Both the **Biscuit, Cake, Chocolate & Confectionery Alliance** and the main flour millers' organisation, the **National Association of British & Irish Millers**, merely accept the content of legislation. Neither uses pesticides directly, and their suppliers are expected to comply with UK law.

The retailer's approach to policy was more positive and,

although there were major differences of emphasis, they tended to show more thorough implementation of their pesticide policies across their operations.

Smaller retailers like **Kwik Save** and **W. M. Low** have neither the purchasing power nor the own-brands which could strengthen their bargaining position with manufacturers. Such lack of market power leaves all small grocers and small supermarkets at the mercy of their suppliers. Many small chains like these also contract out their green grocery, butchery and delicatessen departments to outside operators, which further reduces their control over the food sold.

Safeway, the first of the large multiples to introduce a wide and regular supply of organic food, now sold in all 310 outlets, developed its pesticide policy in the early 1980s. It was influenced by a 'customer-led' approach to the business, in which close monitoring of early trends in new buying patterns and comments from customers provide the signals for action. The total use of pesticides is to be restricted wherever possible and any specific pesticides on the approved list which the company considers suspect or feels are unnecessary are either drastically reduced or banned for growers or suppliers.

Sainsbury's are typical of the large supermarkets which obtain much of their fresh produce direct from growers under contract. This heightens their awareness of both the need for pesticides and also the risks. Company pesticide policy, which began in 1979, acknowledges the 'increased awareness of the effects of pesticides on the environment and the health of the consumer' and concludes that it is 'essential that a clear policy on pesticides is followed'. **Sainsbury's** policy sets out a firm line on pesticides, with codes of practice for suppliers in all departments, not only fresh produce. Their first code of practice including guidance on pesticides was produced in 1982. Pesticides are to be used only 'where they are deemed necessary', and then methods of reducing the sprayed volume are to be followed where possible. Growers are expected to follow all new technologies for reducing pesticide use, including biological control, which 'must be utilised to the fullest extent'. Varieties with built-in resistance are encouraged. Post-harvest chemicals, which are the cause of most food residues, are identified for particular attention.

The growers are also expected to maintain detailed records of pesticide use, to monitor their own residues, and to supply the data to **Sainsbury's**. Growers are bound to follow UK legislation on the approved pesticides and methods for use. These conditions are included in suppliers' contracts and may be added to where the company feels additional steps are necessary.

Tesco also have a strong commitment to reducing pesticide use and residues. Their policy started in 1984 with the introduction of detailed residue monitoring to which were added, in 1986, codes of practice and efforts to improve crop production methods. Like other retailers, **Tesco** were struck with the wide and possibly excessive range of pesticides approved for use on some food crops. For instance, there are 20 pesticides against aphids and 18 against caterpillars approved for use on cabbage in the UK. A **Tesco** code practice stipulates that 'all crop treatments carried out in the UK must adhere to the conditions laid down in current product approvals. All produce moving in trade in the UK must comply with current Maximum Residue Limits [*sic*] as specified by law and suppliers must take all the necessary steps to ensure due compliance. Tesco Stores Limited requires that comprehensive and accurate records are maintained in respect of all chemical treatments applied to produce supplied to the Company and discharges this responsibility to the supplier of the product.'

Waitrose have a similar policy on pesticides, with growers and suppliers having to conduct their own residue testing, which is backed up with the company's own testing programme. **Waitrose** introduced organic fruit and vegetables in 1985 and are also working to reduce pesticide use through either farm management methods or restricting the number of allowable pesticides. Both options reduce pesticide use by encouraging alternatives to chemicals. They are not yet as involved in research and the development of alternatives to pesticides as are **Sainsbury's**, **Tesco**, and **Safeway**. However, **Waitrose** adopt UK legislation and take steps to ensure that all produce sold conforms to either UK MRLs or others specified by the company. They have recently adopted a total ban on tecnazene on all potatoes supplied to their stores.

The remaining three multiples represented gave less detailed information on their policies. The **CWS (Co-Op)** maintain that they have had an overall quality statement since 1884, covering all food sold, but without specific reference to pesticides. However, they do normally include in all specifications to suppliers of fresh produce that UK legislation on pesticides must be followed. **Marks & Spencer** have had a policy on pesticides since the 1960s, but it was based on accepting relevant law, which had a negligible effect on pesticide use before 1985. Detailed specifications are given to growers and suppliers, but while these do specify crop varieties and husbandry methods, there are no general restrictions on pesticides included apart from conforming with relevant law.

How the food industry is adapting to MRLs

During the interviews that followed receipt of the questionnaire, there was a common concern with how the MRLs were interpreted by consumers. Many in the industry felt that consumers had misunderstood the meaning of MRLs and were incorrectly using them as some proxy for a safety limit. As explained on page 109, the MRLs were created to give the maximum amount of residue *expected in produce* when pesticides are used on either the growing crop or the harvest *according to good agricultural practice.* However, the maximum amount of residue, or MRL, is set in full knowledge of the toxicity of the chemical and the amount of each crop, and hence residues, that are likely to be eaten in the average diet. So it is reasonable to consider that an excess MRL may be a risk to health.

The underlying concern in the industry lies not in the possibility of an excess MRL being seen as a health risk, but that any level of residue less than a MRL will also be seen as a risk, even if only a proportional one. For this reason, many in the food industry would prefer to see no measures of health risk attached to residues. This may explain the industry's enthusiasm for UK legislation on MRLs to be interpreted only as guidance for the use of pesticides in practice.

The new Food Safety Act, 1990, has changed the whole industry's attitudes to health risks and hence to MRLs. The requirement for any seller of food to show 'due diligence' in ensuring their food is safe and fit for human consumption appears to give more weight to MRLs. Although the Act is unclear on what 'due diligence' consists of, the MAFF have advised that exceeding an MRL may be taken by the courts as making a food unfit or even injurious to health (see page 125). It has also been suggested that EC MRLs may carry the same weight in court cases involving pesticides without a statutory UK MRL.

This confusion over the meaning of MRLs is not helped by the legislation. Neither the Maximum Residue Levels Regulations (see page 109) nor any of the related legislation makes any mention of the ADI (Acceptable Daily Intake), which is an internationally accepted measure of health risk. Nor are any other measures of health risk mentioned. For this reason, and because health *is* taken into account in setting the MRLs, it is reasonable to assume some unacceptable risk to health is likely where MRLs are exceeded. (See page 121.) However, it is important that this confusion in the legislation be clarified urgently. The industry feels that clarification must wait until the development of case law. This can be a lengthy and not always useful procedure. Far better for the MAFF to tackle the problem and introduce the necessary amending regulations or legislation. At the very least, some measurement of health risk is essential for the Food Safety Act to be enforced effectively. The ADI alone is not enough because it relates only to the total diet. All that is necessary is an acknowledgement of the health component of the MRL, stating that excesses render a product 'liable to be injurious to health'.

The food manufacturers' replies to the questionnaire were generally legalistic and confused in their attitudes to MRLs. They apparently saw the MRLs more as a production constraint than a health safeguard, and wanted their application and interpretation so limited. There was a commonly expressed view that MRLs were not to be used in any way as safety limits. Despite the small number of UK pesticides with MRLs, it was also said that the only MRLs which farmers or manufacturers had to pay attention to were those in UK law. The **Biscuit, Cake, Chocolate & Confectionery Alliance** referred to the CAC MRLs as guidance levels, 'but it should be understood that there is no obligation to meet them nor is there necessarily any threat to health from a residue level which may from time to time exceed it'. They also noted that, although there are five pesticides with CAC MRLs registered for use in cocoa production (dichlorvos, fenitrothion, fentin, lindane, and delta methrin), 'cocoa contributes so little to the UK diet and so it is not surprising that UK MRLs have not been established'. The **FDF** reply to the question on whether manufacturers found that residues in ingredients made it difficult to keep residues low in the final product said: 'The only legal constraint are the relevant MRLs.' As the law now stands, MRLs apply only to farm products, and no manufactured foods are covered.

The **Brewers' Society** informed its members in January 1990 that retailers had been asking for assurances that beer complied with the UK MRLs. The Society's position then was that 'there is no legal requirement on food and drink manufacturers to check that their raw materials comply with MRLs. However, in the interests of the customer [i.e. the retailers], brewing companies will wish to assure themselves that their supplies are wholesome.' Brewers have told their hop suppliers to restrict EBDC fungicides and are now taking steps to ensure that their malting barley supplies do not exceed MRLs.

No manufacturer or its trade organisation suggested that the UK should increase the number of pesticides with UK MRLs from 62 to include the full EC list, or even the more extensive CAC list of the FAO. Nor was there any suggestion that EC MRLs should be given even a guidance status in the UK until the UK list is expanded. The approach of the retailers was very different.

All the retailers said they followed both UK MRLs and additional MRLs: most followed both the EC and CAC MRLs but **Waitrose** also took those of the US into account. These sources were followed for pesticides not on the limited list of those with UK MRLs, and usually the lowest MRL was followed. Discussion on MRLs with the retailers was completely different from that with manufacturers. Most wanted a MRL for every pesticide to provide a common reference point for monitoring pesticide use. There was little sign of the manufacturers' eagerness to stick to the scant content of relevant UK law. The retailers were more concerned with what pesticides were needed and with safety levels than with sticking to either the list of UK approved chemicals or the UK MRLs. A produce buyer in **Sainsbury's** said they stopped their suppliers using any alar on apples 'because we saw no need for it'.

Specifications for growers or suppliers used by **Tesco, Waitrose**, **Safeway**, and **Sainsbury's** often include the need for non-UK MRLs to be followed where there were no UK values. The retailers have demonstrated their intentions to reduce unnecessary pesticide use by imposing lower-than-UK MRLs on several pesticide/crop combinations.

Some supermarket restrictions which exceed UK MRLs

Retailer	Pesticide/crop	Restriction
CWS (Co-Op)	alar/apples and pears	not allowed
Marks & Spencer	alar/apples and pears	not allowed
	tecnazene/potatoes	MRL 0
Safeway	alar/apples and pears	not allowed
	tecnazene/potatoes	not allowed
	lemons and oranges	no post-waxing or post-harvest treatments
Sainsbury's	alar/apples and pears	not allowed
	tecnazene/potatoes	not allowed
	maleic hydrazide*/ potatoes	not allowed
	lemons*, oranges*, and apples	no waxing or post-harvest treatments
Tesco	dithiocarbamates/ lettuce	<UK MRL
Waitrose	alar/apples and pears	not allowed
	tecnazene/potatoes	not allowed
	maleic hydrazide/ potatoes	not allowed
	dimethoate/lettuce	MRL 1.0 p.p.m.

* These restrictions apply only to special lines labelled 'free from post-harvest treatments' or similar.

The industry's own residue monitoring

The manufacturers have hitherto relied on their suppliers to comply with pesticide regulations, often requesting a written warranty that only good agricultural practices have been followed. However, warranties are no longer an accepted defence under the new Food Safety Act, and monitoring is considered a necessary part of 'due diligence'.

Few survey replies from manufacturers placed much weight on their monitoring programmes. The **FDF** outlined the difficulties facing members. 'Due to the wide range of analyses which might be performed and the sophisticated techniques required, frequency of testing must depend upon need and resources be directed to identifying potential problem areas and concentrating analyses on monitoring them.' This inevitably means many manufacturers would test for as few pesticides as possible and test as required rather than maintaining a routine monitoring system.

The **Brewers' Society** and flour millers conduct regular tests for residues of chemicals they expect have been used. The **National Association of British & Irish Millers** organises a collaborative monitoring where the 45 member companies, accounting for over 99% of UK milled flour, send samples to the national Flour Milling and Baking Research Association for residue testing. An annual report is published. Samples from imported wheat as well as UK crops are tested separately for a range of pesticides including organophosphates, organochlorines, and fumigants such as inorganic bromide. Data submitted from the 1989/90 survey show a tendency for UK-grown wheats to have higher residues than wheat from any other country but that, with the exception of lindane on some samples, residues of all pesticides were below UK MRLs.

The retailers' residue monitoring is more systematic and intensive. Most require suppliers of fresh produce to carry out residue monitoring to agreed standards, often in approved commercial laboratories, and to an agreed timetable. All crops are routinely tested each season, according to the growing cycle. High-risk, high-volume lines like lettuce might be tested every two weeks, whereas lower-risk, lower-volume lines like swedes might need to be tested only once a season. **Safeway, Sains-bury's, Tesco**, and **Waitrose** all require suppliers to conduct tests and submit results. Follow-up tests may be required if MRLs are exceeded or if there is reason to suspect unacceptable residues. **Marks & Spencer** do not require all their suppliers to monitor for residues, but they say they take samples on delivery and send them to commercial laboratories for testing.

The intensity of testing is higher and more systematic than that of the WPPR in MAFF. **Safeway, Sainsbury's, Tesco**, and **Wait-rose** require every grower's or supplier's produce to be tested at least once a year, and, depending on the crop, many lines are required to be tested fortnightly. **Marks & Spencer** and the **Co-Op** take samples from the total delivery of each line rather than from each single supplier, but at least they say they are testing every line at a minimum of once a year.

Most retailers do not conduct their own testing but arrange for samples to be sent to reputable commercial laboratories like Restec in Worcestershire or the MAFF laboratories in Harpenden. **Safeway** and **Waitrose** have their own laboratories, allowing a reduction in the time taken to obtain test results. **Safeway** said they get their results in three to four days rather than the three to four weeks it took when they used external laboratories. This is very important as it allows perishable produce like strawberries to be sampled in the field and approved before it reaches the supermarket shelves.

Imported supplies are usually tested before shipping, and often the UK supplier is required to take additional samples for analy-sis. **Marks & Spencer** cited a Chilean source of grapes where the grower applies pesticides according to the specific regulations of each importing country and where quality control is both of a high standard and protected by tight security arrangements. **Marks & Spencer** technologists inspect both the growing of the grapes and the packing facilities, and accept warranties from the supplier.

The retailers with detailed monitoring arrangements also have strict procedures for following up excessive residues. Suppliers must keep records of all samples tested, and each record contains data on the grower which enables further checks on pesticide use. Normally, any excess residue is investigated to discover the cause, and subsequent deliveries are more intensely monitored to ensure the problem has been solved. Failure to correct them usually means the supplier's contract is not renewed.

The residue data that were presented during the *Poisoned Harvest* survey are impressive because of the high proportion of samples in which residues were reported to be below MRLs. The respondent's word was accepted on their residue data. All retailers reported some samples which exceeded MRLs. Most reported detecting excesses of dimethoate or EBDC in lettuce on at least one occasion. Tecnazene in potatoes and alar in apples have also been reported. Other excesses that have been detected include alar in apple juice and copper sulphate on grapes. In all cases only one or two samples with residues above the MRL were reported.

Individual cases of excesses had been followed up. One was on a test crop of lettuce which was ploughed in once the residues were discovered. Others were discussed with the growers and were acceptable on subsequent retesting. Once a grower or supplier had found an excess MRL, they were normally required to carry out special tests to ensure the problem had been corrected. One retailer found that tecnazene was not being applied according to label specifications by several growers delivering to a potato supplier. The response was a special set of directives for the growers concerned, with additional testing at the growers' expense. Subsequent tests showed tecnazene fell to below the MRLs.

The retailers' residue data suggest much lower residues than in the rest of the food supply as measured by the MAFF's own residue monitoring. Not only do the retailers have clout through their market power, they have also been active in encouraging growers to use less pesticide. The effect of this should not be undervalued. Growers with contracts to supply the large supermarket chains have a guaranteed outlet for all their production at agreed prices. As the retailers grow more sensitive to their customers' interest in pesticides, they are building more checks and controls into contracts with their suppliers.

Tesco, for example, has initiated seminars with the growers of crops like potatoes, tomatoes, cucumbers, and round lettuce. Emphasis is on finding methods of cultivation and storage which will reduce residues. By 1991, all **Tesco** peppers will be grown with biological control rather than with chemical pesticides. Potato farmers are being encouraged and assisted towards long-term cold-storage methods which will remove the need for tecnazene and other storage pesticides. **Sainsbury's** are also supporting research into better cold storage of potatoes. In addition, they are encouraging apple growers to convert to natural, vegetable waxes instead of mineral types, and they are supporting research programmes on biological controls for mushrooms and other vegetables, and on computer models which can give early warnings of disease and so help reduce unnecessary pesticide applications. Where pesticides are needed, **Sainsbury's** are funding research into new methods of spray application which use lower volumes and reduce the drifting of droplets.

On a larger scale but in the same direction is **Safeway**'s joint research venture into organic farming methods in Scotland. The research is based on the 300-acre Jamesfield Farm in Perthshire and will look at the economics and improved cultural techniques, with the aim of reducing the price of organic foods and improving the regularity of supply. The project is a collaboration between the EC, the Scottish Development Agency, the Edinburgh School of Agriculture, and **Safeway**.

Labelling

Labelling foods with pesticide information is difficult, but the same was said of nutrition labelling in 1980. All respondents to the survey were wary of the difficulties, and none had plans to introduce such information.

Responses ranged from the dismissive to reasoned resistance. The **Biscuit, Cake, Chocolate & Confectionery Alliance** declared that residues in chocolate were below MRLs, so, 'since nobody was at risk there is no benefit to be gained by the consumer from the declaration of pesticides on labels. Indeed it would be non-sensical and misleading, bearing in mind that the sensitive analytical techniques available today are able to detect traces in virtually all foods.'

The **FDF** preferred the option of labelling only organic foods where no pesticides had been used. They saw practical difficulties in labelling foods with all the pesticides that might be applied, especially for manufacturers who combine different ingredients in one product. There could also be problems when one ingredient came from a number of sources because different pesticides could be used at each source. **Sainsbury's** and **Marks & Spencer** felt that detailed labelling could bring pressure from campaigning groups wanting particular chemicals banned.

A third alternative may be of more use to consumers and have fewer practical problems. With growing concern focusing on individual pesticides rather than on the avoidance of all pesticides, labels could usefully declare when a particular pesticide has *not* been applied. Tecnazene-free potatoes, alar-free apples, EBDC-free lettuce and cucumbers, for example. As more retailers and growers are avoiding particular chemicals, this form of labelling is both useful information and practical. The main objection from the retailers is the fear that consumers will see a label, 'tecnazene-free potatoes', on one shelf and will treat all unlabelled potatoes as suspect. This same fear delayed the introduction of own-brand nutritional labelling by supermarkets and was proved groundless as soon as the labels were introduced. With several supermarkets now selling potatoes 'free of post-harvest treatments', and waxless oranges, lemons, and apples, can this useful labelling be far away? Which retailer will be the first to meet this consumer need?

Publishing residue data

Releasing information to consumers still doesn't come naturally. Manufacturers and their organisations were unhappy with the thought of making public their residue data but suggested consumers could write to individual companies. Whether this would elicit only assurances like 'none but the safest of chemicals and procedures are used on our products' will have to be put to the test by individual consumers.

Although retailers do not yet publish their results, they should be open to encouragement. They are understandably concerned about confidentiality, but, while these concerns were emphasised during interviews, none would stand examination as an excuse to withhold data on residue levels. The retailers felt, first, that residue data on a grower or branded supplier could have unnecessary, and even unjustified, commercial consequences if published. Many residue excesses are accidental or may not be the fault of the grower. Second, retailers fear that if they have one more excess MRL detected a year than their competitors, consumers will switch purchasing habits or campaigners will bear down heavily.

Retailers were very coy about the results of their residue testing being made public in this book. Such coyness is surely counter-productive. With food additives and nutritional labelling in the 1980s, it was the lack of information which led consumers to react decisively, if excessively on occasions. Retailers should be contributing to the education and information of their customers by explaining their residue policies and making more residue-testing data available. It would be sufficient to publish summary data, listing numbers of tests, average residues, and number of samples exceeding MRLs. Were this put in the context of re-tailers' action to reduce pesticide usage and, in particular, the action taken on any cases of excesses, it would be both useful to consumers and of a genuine public-relations benefit.

Marks & Spencer felt that both labelling and the publishing of residue data would not be appreciated without customer educa-tion. **Tesco**, **Safeway**, and **Sainsbury's** have already established reputations for the quality of their in-store education materials on nutrition and additives, among other subjects. Pesticides, especially with the significant actions being taken by retailers, should also be a subject of customer education.

There are moves which may encourage these changes. The Campden Research Association is preparing a pooled data base on residue data, with information collected from across the food sectors. The Minister of Agriculture has also invited the food industry to share its data with the MAFF to expand the data available for its national monitoring of residues.

Supermarkets as gatekeepers

The survey suggests that the large supermarkets are fulfilling the role consumers now expect of the government. They apply UK law but then go further in restricting pesticides they feel are either unsafe or unnecessary. They have organised monitoring systems that ensure that every type of fresh produce is tested for residues. Tests are carried out either by the supplier or by the supermarket. And, most important, they are using the contracts with their suppliers to enforce both the UK law and their own strict quality requirements.

The best of the large supermarkets are the gatekeepers for consumers. If we relied solely on government monitoring and enforcement, such as it is, any contaminated food would be sold and consumed long before any excess residues could be detected. The supermarkets protect consumers from excess residues by ensuring that unsafe food is highly unlikely to enter their warehouses, much less the shopping basket.

Four supermarkets stood out as the most effective in reducing pesticide risk. The factors considered include contracts with suppliers and specifications, monitoring and quality control covering residues, research, and other initiatives to reduce pesticide use and find alternative pest controls. The top four supermarkets are:

Safeway
Sainsbury's
Tesco
Waitrose

Two other supermarkets also showed significant initiative in reducing residues on their produce but appeared not to follow the same high standards as the top four on at least two criteria. They are:

Marks & Spencer
CWS (Co-Op)

Marks & Spencer have less comprehensive monitoring than the top four. They analyse only a sample from each of their lines rather than from each supplier's produce. They also showed less concern to identify and reduce, on criteria like unacceptable health risk or unjustified use, those pesticides not on the UK approved list. The **CWS (Co-Op)** relied on an overall quality statement in product specifications but did not include specific restrictions on pesticides apart from following UK law. Monitoring was not as intensive as that practised by the top four. The **CWS** own farms which supply some of their produce and are developing organic methods which will increase their own supply of fresh vegetables.

Consumers concerned about pesticide residues can buy with far greater confidence from these supermarkets than from any other high-street source – barrows, greengrocers, mixed corner shops. Unless they are certified organic, 'pick-your-own' farms and farm shops are rarely any better. Non-supermarket sellers are not necessarily less careful; rather, the smaller the chain or shop, the less likely it is to have the resources to carry out residue testing or control its suppliers. These additional controls would not be necessary were the government controls as effective as the publicity from the MAFF would have consumers believe. In fact, much of the produce which the supermarkets reject because of high residue levels ends up in high-street greengrocers and barrows.

The large supermarkets may offer consumers additional benefits in the future. Already they have taken a strong stand on safety issues like food irradiation and *Salmonella*. While they are showing concern for the inadequacies in current legislation on pesticides, they are in a good position to lobby for necessary improvements. Their investment in research on alternatives and in agricultural systems like organic farming should increase their interest in encouraging both future food policies and legislation which supports these trends. It seems reasonable to forecast that, unless the supermarkets change their priorities, there should be a continued fall in the level of unnecessary pesticides and in residues of *all* pesticides in the foods sold by supermarkets.

The influence of supermarkets on farmers and manufacturers maintains a strong consumer influence on the food system. At present, most supermarkets have little investment in manufacturing or agriculture. They have little interest in who supplies their food so long as their high quality standards are met. They make their money from selling food, and keeping the consumer happy is more important than keeping the supplier happy. While this relationship remains, consumer influence on supermarkets is strong. It is important that consumers recognise this power and know how to use it.

Individuals have significant power through their choice of whose trolley to push up and down the aisles. But there are additional influences open to consumers. Like MPs, supermarket managers respond to customers' letters. Don't write only with complaints, though. Let them know when they please you as well. Supermarkets are cautious innovators and will test-market new products or ways of selling. Their most reliable measure of success is the speed at which something moves off the shelves. But your letters can encourage them to introduce new ideas and also to understand why something did or did not move quickly enough. The introduction of pesticide labelling and in-store customer information on pesticides could be speeded up by writing to your supermarket saying you want them. Even something so delicate as publication of their residue-monitoring data could be encouraged by customer persuasion.

There are criticisms to be made of supermarkets, and plenty of people argue that they work as much against the consumer's interests as for them through the potential abuses of their near-monopoly power. These are worthy arguments and deserve debate but are not relevant to the issue of pesticides. On food safety generally, and most dramatically with pesticides, the six large supermarkets identified here are a major influence in making pesticides safer.

One possible danger for the future bears contemplation if only to avoid too much complacency about the community spirit of supermarkets. Were the large multiples to begin investing significantly in the manufacture or agricultural production of food, their interests and loyalties could become severely divided. At present, their only business interest is in the food they sell. They can apply rigid quality standards on their suppliers, whose profits may suffer under the severe competition created by the power of the large retailers. Any significant ownership of manufacturing or farming would mean much greater interest in the profits of those industries and this could lead to compromises on the quality of the food produced.

The supermarkets are normally reticent when it comes to publicising their activities. Their position is difficult because they cannot be seen to be criticising either their supply industries or government departments. However, it may be argued that, since they have embarked on many initiatives which are having an impact on the environment and on health, they should be more open. With their activities on pesticides, the advantages are clear. For their customers, it would be helpful to know of research initiatives aimed at reducing the use of pesticides, of particular pesticides being restricted, and of the trends in residue levels. Public awareness could be improved, and there is little doubt that as environmental awareness takes greater hold in pressure groups, government, and industry, extra efforts to inform and encourage the public will speed adaption and help give it coherence. The supermarkets are already moving this way on recycling. They are arranging for more recyclable packaging and also for facilities for customers to deposit materials with potential for life beyond the rubbish tip.

Pesticides are ideally suited to a similar response from the supermarkets. More food is bought through stores of the six supermarkets than through all other outlets combined. Supermarkets work for customer loyalty and trust and should involve customers in their own efforts on pesticides. This can only increase customer support for the supermarkets' initiatives and contribute to a better informed population.

Conclusion

The range of cautious but practical measures being taken by the food industry to protect both the environment and consumers from pesticides is impressive. The extent of restrictions placed on pesticide use by some manufacturers, the details on some monitoring systems, and the efforts of the National Farmers' Union and others to encourage alternative pest controls are at least encouraging.

The fact that the initiative is being taken by the supermarkets is welcome but should not deflect attention from the government's responsibility to ensure a safe and wholesome food supply. All the actions taken by the supermarkets should have been taken by government but were not. Not all groceries are bought through supermarkets, and nearly 25% of all food eaten in the UK comes from catering outlets and never encounters the high standards of the supermarkets. The government must be pressed to ensure that every consumer in the country, wherever they shop, benefits from the same planned reduction in the pesticide content of their food as that now provided by supermarkets.

The government has contributed new legislation which offers the potential for better training of pesticide users and improved control of the specific uses and methods of application. However, it has failed to allocate the resources and develop the systems necessary to ensure that the intentions behind the new legislation become facts of life. It still has not found the resources to review old pesticides and to ensure that only safe pesticides are approved. Official monitoring of pesticides is woefully inadequate and, despite new legislation on residue levels, there is little sign of the legislation designed to protect against pesticide abuse being enforced.

The government has left consumers knowing about the improvements in pesticide controls but has not created any confidence that these benefits are being received. These chemicals are odourless, colourless, and tasteless once applied. Without the creation of effective monitoring and enforcement, consumers can do little but suspect the worst. How are we to identify food which has no illegal residues, to know whether or not our drinking water is contaminated, or to know that children and pets are not at risk in the local playground? Consumers cannot conduct their own analyses. They must have the protection of a system that is *seen* to keep residues to a minimum and to deter use of illegal chemicals. While the government continues to leave doubts over the safety of foods, the supermarkets offer an immediate opportunity for consumers to buy with confidence.

They have shown that the public wants fewer pesticides used and smaller residues, and, most important, that these reductions are possible. They have also shown that effective controls on residues are possible, that there are alternatives to the more harmful chemicals in use if only the effort is made to look for them. The government needs to adopt the same approach and apply it not only to the whole of the food supply, but also to the use and control of all pesticides in the environment.

The head offices of the six supermarkets with commendable records on pesticides are:

Public Relations Department
Co-operative Wholesale Society Ltd
New Century House
Manchester M60 4ES

Edward Thomson FIEH
Director for the Environment
Safeway plc
6 Millington Road
Hayes, Middlesex UB3 4AY

Public Affairs
Marks & Spencer
Michael House
Baker Street
London W1A 1DN

Public Relations Department
J Sainsbury plc
Stamford House
Stamford Street
London SE1 9LL

Consumer Relations Department
Tesco
Delamere Road
Cheshunt
Waltham Cross, Herts. EN8 9SL

Customer Services
Waitrose Limited
Doncastle Road
Bracknell, Berks. RG12 4YA

9 Conclusions

★ Consumers should have more involvement in decisions on the safety and regulation of pesticides.

★ Pesticide approval should be based on the assumption that pesticides are harmful until proven safe.

★ Farmers need more help to escape from the pesticide treadmill.

★ Government should devote greater resources to biological pesticide controls and to management systems which require less use of existing chemical pesticides.

★ Monitoring of pesticide usage and of residues must be improved.

★ All safety-testing information submitted for pesticide approval or reviews must be made freely available for public assessment.

★ Labelling schemes for products and foods containing pesticides should be explored with consumer groups and industry.

Public awareness of pesticide use and abuse has increased rapidly. A diet laced with pesticides and an environment awash with them are no longer acceptable. They probably never were; it only appeared so because there was little complaint from a public ignorant of where pesticides were used or what harm they could do. It has been only in the last decade that pesticide use and safety have had wide and regular exposure in the UK media.

Very few consumers reject all pesticides, but their reaction must be understood. The secrecy on safety data, the lack of product labelling, and regular, unsubstantiated assurances that suspect pesticides already in use are 'harmless to humans' have encouraged overreaction and disquiet. However, most consumers are changing their compliant acceptance of pesticides with a determination and logic that must be heard sooner than later.

It is time to realise that no amount of 'scientific assurance' from experts or tendentious pleadings about pesticides being essential to feeding the world will resolve the public disquiet. The consumers simply do not want pesticides about which they have little information and even less say. The clarity of this message still escapes many in industry and government, but its effects are unambiguous in environmental campaigns and in the actions of many supermarkets. Consumers accept the need for some pesticides and accept that there will always be some risk. But they will no longer be kept out of discussion by secrecy or satisfied by too eager assurances of safety. Pesticides and their use must be treated differently in future. Priority must be given to developing chemical-free alternatives and to ways of reducing the use of necessary chemicals.

Need for proof of safety, not harm

The repeated withdrawal of pesticides which, until the day their unsuccessful review was announced, had been protected by Ministers' assurances of total safety does not encourage confidence in the approvals process. Better safety tests, which the older pesticides often fail, are available. These developments in technology are welcome, but they highlight an area where fundamental change is necessary. Once pesticides have received approval, they are deemed safe and secure unless and until new evidence is sufficient to show them to be harmful. If shown to be harmful enough, they can be restricted or withdrawn. The difficulty often lies in convincing the Advisory Committee on Pesticides (ACP) that evidence of harm is 'conclusive', or in deciding what is 'harmful enough'. Given past errors like those made with DDT and the increasing numbers of approved, that is, 'safe', chemicals that are being withdrawn subsequently on safety grounds, a reversal of emphasis is required. First, the criteria of safety must be more stringent before approval is given, and, second, when any approved chemical comes under suspicion, the onus must be on proving its safety and not its harm. Both harm and safety are difficult to assess, but the balance should shift to the most stringent criteria for safety. Questions of harmful effects should lead the regulating authorities to demand tougher tests of safety to justify continued use.

A fundamental improvement would be the introduction of an appeal process. This would aim to allow other scientists to examine submitted test data and also to review the ACP's decisions. This is analogous to the appeals process in law but need not be so formal or protracted.

Farmers dance to the agrochemical fiddler

Spraying chemicals for prosperous farming is like fighting for peace. But farmers are now trapped on a pesticide treadmill. New crop varieties are genetically weaker than their forebears and require more protection from weed competition and damage from insects or fungi. Intensive farming of animals has increased the number of parasites and diseases. The worst enemy of animals is overcrowding in pasture, pig shed, or battery house, all of which increase disease spread. The cost is dependence on pesticides against internal and external parasites, and infectious diseases.

As resistance to chemicals increases, farmers are compelled to use more pesticides more often. An added pressure comes from the food markets, where consumers have been trained to accept more and more perfection in appearance rather than in taste, texture, or 'goodness'. Blemish-free food is pesticide-dosed food. It is ironic that the higher-quality, sanitised, cosmetically tantalising fruit and vegetables that consumers are told they demand exist only because, from seed to supermarket shelves, crops are doused in a series of life-killing chemicals. Can it be long before supermarkets will be compelled to sell lettuce in sealed, biodegradable, see-through bags that contain at least one live aphid?

Any lettuce without its live aphid will be left on the shelf, whatever the 'sell by' date.

There are welcome changes in the use of pesticides, but both the marketing strategies of agrochemical companies and the economics of farming are keeping the pesticide treadmill seductively spinning. Farmers can change only as fast as new techniques are developed and fitted within the farming system, backed up by pricing packages which encourage change in well-established habits. Farmers have been encouraged by successive governments in the 1970s and '80s to produce the highest yields possible from their land. They were paid for yield and not for nutritional or health quality. The dominant criteria of quality applied to fresh fruit and vegetables are size, evenness, and freedom from blemish, all of which depend on pesticides. The whole farming system – including the machinery and buildings, the choice of seed or animal, and the inputs like fertiliser and pesticides – has been dictated by government policy in both the UK and the EC. All parts of the system are interrelated. Major changes require both support from the suppliers of technology and the right economic incentives from the government.

This demands a major policy shift. At the least, there should be strong incentives for farmers to comply with existing legislation on pesticide use, backed by better monitoring and stiff penalties to deter the less conscientious. But there should also be incentives for farmers to adopt the more environment-sustaining farming technologies. Integrated pest management (IPM) and biological controls are important and fit comfortably within a shift from intensive farming to more extensive systems. The recent policy of set aside, which pays farmers to take land out of producing crops in EC surplus, could be used more advantageously. Instead of this land going to non-farming uses, farmers could be paid to farm the same land less intensively. This means utilising all the farmland resource, and keeping it in 'good heart' so it can be readily put into higher output if the future demands. It is also an opportunity to develop and introduce organic and other extensive or low-input systems which can produce much of the UK's food needs with less environmental damage.

These low-input systems have been discriminated against by policy for so long that their introduction should be supported with positive discrimination. UK agricultural policy has consisted of favoured crops and systems of production to direct production and support farm incomes. The positive support for less pesticide-dependent systems requires no radical policy change, just the commitment.

Alternative pesticides

New techniques are being researched and developed. IPM is a grand name for efficient husbandry. Instead of spraying crops or treating animals according to a timetable, just in case any one of several possible pests appears, the wise farmer now sprays only when the pest is likely to cause trouble. Knowledge of pest life cycles helps predict when optimum conditions for pests like aphids or leaf moulds occur, and protective spraying can be arranged as necessary. The result is fewer sprayings and less spray being used on crops.

Another approach is to divert plant breeders from their obsession with increasing yields. A new objective is rebuilding plants' genetic resistance against natural pests. The logic is impeccable. Instead of spraying pesticides until the pests become resistant to the chemicals, breed plants to be resistant to the pests and to need little or no help from sprays. There are great implications for food production in the Third World, where resistant varieties could save much loss of production while also saving lives because less pesticide would be used than under the normally uncontrolled conditions. David Sainsbury, of the supermarket family, allocated £2 million from his private trust in 1990 to build a new laboratory dedicated to research in plant resistance. There will be 45 scientists at the new Norwich laboratories with £1.4 million to spend each year for the next 10 years.

Cosmetic pesticides, like tecnazene, which are used to inhibit potato sprouting, are also receiving the attention of plant breeders. Pressure from supermarkets to drop unnecessary pesticides may have given the impetus to breed sprouting-resistant varieties. The Plant Breeding Institute in Cambridge has produced a new potato variety known affectionately as Shula. It cooks like a King Edward but has a long natural dormancy which reduces the need for tecnazene.

Biological controls are set to be the flavour of the decade. These tackle pesticides by restoring natural predators among pest populations. This may be as simple as making sure ladybirds can multiply on crops where aphid populations are building up. More often, scientists are looking for specific organisms, usually parasites, which can be bred in large numbers and released on crops where specific target pests are problems. This is most effective in glasshouses, where there are large populations of plants under a controlled environment.

There are many types of biological control available. One of the earliest to be exploited is *Bacillus thuringiensis*, which is a bacterium used against caterpillars. There are fungi, like *Verticillium lecanii* which is used against thrips and whitefly; parasitic wasps such as *Dacnusa sirbirica* which is used against tomato and chrysanthemum leaf miners; and mites such as *Phytoseiulus persimilis* which is used against other mites like the two-spotted mite. There is even a friendly nematode (*Sternernema bibionis*) which is used to parasitise the vine weevil.

These 'friendly' organisms are of the same types which supply the important agricultural pests – bacteria, fungi, wasps, and nematodes. The fact that there are useful members as well as pests in each group is a reminder of the interdependence within nature. Many chemical pesticides cut through nature's organisms like a scythe, killing both pests and beneficial organisms indiscriminately. This destroys the chains of interdependence among organisms and can lead to new pests emerging as their natural predators are killed off. As the predators are killed, more chemicals are needed to control pests. This is an important cause of the farmer's pesticide treadmill. Agrochemical companies are trying to develop more specific pesticides to kill only the target organisms.

Information gives freedom of choice

There can be little doubt that the crisis in UK pesticide policy has escaped radical overhaul for so long only through the power of secrecy. Consumers are deprived of information on many fronts, but two in particular need urgent attention. First is confidentiality on safety test data and ACP decisions. Data submitted to the government on the nature and size of risks to health or environment are withheld, as are the details of government assessments of these data. Safety data on the same chemicals, usually provided by the same agrochemical companies, are often available to the public in North America. This demolishes the common justification that commercial secrecy must be protected in the UK. There can be no acceptable argument for keeping data on a pesticide's safety or efficacy confidential. Pesticides are poisons which are designed to be spread throughout every environmental niche and house skirting board. Consumers have a right to know what a pesticide is meant to do and to know the health risks. An important social benefit would be that publication would allow scientific scrutiny of a company's own test data and public debate on the basis for individual approvals.

Companies developing new chemicals are entitled to have data on the composition of active ingredients and formulation of pesticides withheld until they can place them on the market. However, there is no argument for commercial secrecy over data submitted 5, 10, or 15 years ago. Where protection is at issue, there are patent laws which, if they can be adapted to cover the genetic material in seeds, must be capable of covering synthesised chemicals. Both companies and government officials argue that test data submitted for a pesticide's approval remain the property of the company. The government has the data on loan from the company for a specific purpose. It cannot be passed on without the permission of the owner, the company. Just as evidence in courts of law and data in the annual reports and accounts of all companies are made public, so should the equally important information on the testing and assessment of pesticides be made public. The solution is remarkably simple. It can be made a requirement of any submission for approval that data provided be in the public domain. Any company not wishing their data to be available to the public need not apply for approvals.

Second is the lack of information on what pesticides are used where. It is unreasonable to expect water suppliers to notify consumers of all individual pesticide contaminants. Accidental contamination is difficult to predict. However, much consumer contact with pesticides results from deliberate, planned, and usually recorded use of chemicals. Most pesticides are used without public knowledge or consent. There is good reason for consumer access to information on pesticides used in parks or verges, around places of work, and on foods. There can be no freedom of choice to avoid a source of these potentially dangerous chemicals without knowledge of what is sprayed where. Parks could display notices when staff are applying pesticides; householders and the public could be notified when weeds or vermin are to be treated in streets and verges or houses; water companies could notify customers when pesticides are added to treat the water; and cut flowers and pot plants could be labelled with the pesticides applied during cultivation.

More difficult but no less important is the question of food labelling. Farmers should be required to make known the pesticides used on crops, especially any post-harvest treatments. The law covering food labelling should be amended to include certain pesticides, for example, post-harvest treatments on fruit and vegetables. At present, manufactured foods must carry a list of ingredients. An ingredient is legally defined as 'any substance, including any additive and any constituent of a compound ingredient, which is used in the preparation of a food and *which is still present in the finished product, even if in an altered form*' (my italics). An additive is defined as 'any substance, not commonly regarded or used as food, *which is added to, or used in or on, food at any stage to affect its keeping qualities, texture, consistency, appearance, taste, odour, alkalinity or acidity* . . .'[1] (my italics). Pesticides could be included neatly within these definitions but are not mentioned in the labelling regulations. However, fresh fruit and vegetables are specifically exempt from the need to label any ingredients. These two anomalies could be removed.

Such radical changes in the provision of information on pesticides could have significant benefits to the community. Wider scrutiny of test data would both improve the quality of data submitted and reduce the chances of unsafe chemicals being approved at all, or at least reduce the time before they are identified and withdrawn. It would increase public confidence in the approvals and review process of government. It would also discourage the unnecessary uses of pesticides and encourage a more informed consumer response to those uses of pesticides which are both socially useful and 'safe'.

As the experience of food additives has already demonstrated, better consumer information has reduced the unnecessary use of additives and simultaneously increased consumer confidence in manufacturers taking a responsible approach to both choice of additives and informing their customers. Only irresponsible pesticide users need fear an informed public.

Consumers pay and deserve more say

Environmental concerns are widespread among consumers and are leading to action. Well-informed and well-intentioned public pressure from many quarters, not only the pressure groups, has influenced important changes in government policy. The sudden emergence of 'green issues' on the hustings in the last European Parliamentary election and in the UK national elections are good indicators of politicians detecting the winds of change. However, what is said on such occasions and what is done subsequently are not often closely related.

The historic removal of decisions on pesticides from public involvement need not continue. Pressure for change is mounting. Few areas of government policy are so intimately concerned with individual consumers as is the provision of a safe, nutritious, and enjoyable supply of food. Yet few areas are so removed from public involvement in the policies responsible. Both consumers and government need reminding of the extent to which agriculture is dependent on public funding; to which the whole approvals and monitoring process is for the public benefit and not the companies'. The public also pay for the operation and running of the government departments like MAFF and DOE as well as agencies like the Health and Safety Executive, which, between them, oversee the approval, use, and monitoring of pesticides.

The UK government is acutely concerned that the market should decide how resources are used. Pesticides are therefore in need of urgent attention. Not only do consumers pay handsomely for the privileges of a local agriculture, but they also pay for the system responsible for regulating pesticide use. The growing voice for fewer and safer pesticides, and for better checks on the use of pesticides deserves more attention. Consumers pay for pesticide regulation and should have more say.

The big supermarkets, supported by the environmental and other consumer groups, have set the pace in more responsible approaches to pesticide use. They give consumers a direct link to the use and regulation of pesticides in the food industry. Significant changes have occurred already, and, with continuing contact between the supermarkets and their customers, there is little sign of this initiative diminishing. Consumers interested in safer pesticide use should direct their energies towards their supermarkets, making suggestions, asking for information, and welcoming initiatives they agree with.

As pesticide use in agriculture and the food industry becomes more health and environment sensitive, other areas of use will come under the same pressure for change. The tighter controls on pesticide choice and use adopted by the supermarkets will influence government procedures sooner or later. Perhaps there will be a Green Revolution in agriculture after all.

REFERENCES

1. HMG (1984). *The food labelling regulations 1984*, para 2. London: HMSO.

10 Practical Ways to Reduce Your Contact with Pesticides

Much of this book has described pesticides as chemicals which are difficult, if not impossible, for the public to detect once they have been applied. It has also shown how action by government and industry is essential if the health and environmental risks from pesticides are to be reduced.

You may feel frustrated because the ear of government seems so far away that your voice is never heard. The supermarket survey in Chapter 8 should give some encouragement because how and where each of us chooses to spend the household food budget **does influence** the supermarkets, and **they** are influencing industry and the government. However, there are plenty of simple measures which you can take to reduce both your own contact with pesticides and any unnecessary leakage into the environment.

Careful thought should be given to the two broad categories of exposure: voluntary, which includes spraying in your own home or garden, and involuntary, which includes residues in bought foods and contact with chemicals in the environment. The following are helpful tips:

- Use pesticides only when you are convinced they are necessary. (Don't empty a can of aerosol fly spray for a single fly in the kitchen; don't aim to kill every caterpillar or aphid you see, as small numbers do little or no damage and soon attract birds or other insects to harvest them.)

- Make a conscious decision to look for control methods which do not require pesticides. (Ladybird larvae devour vast numbers of aphids; hedgehogs and simple traps control snails and slugs more safely than baits.)

- Always read the label carefully and follow the instructions and any precautions to the letter.

- Check the ingredients of branded pesticides. Using *Poisoned Harvest*, find out what chemical type they fall into and how safe they are. Use only pesticides which contain chemicals you feel comfortable with.

- Apply only enough chemical to do the job and resist applying that 'little bit extra for good measure', as it rarely helps and may cause harm.

- Be strict about storing and applying *all* household and garden chemicals away from food, children, and pets.

- If you use pet flea-collars, don't handle them when fondling the pets. If your children aren't old enough to follow the same advice, tackle the fleas another way.

- Observe any instructions on the labels of pesticides used on garden or allotment food crops about the minimum period between spraying and harvesting.

- Wash all fruits and vegetables, especially leaf vegetables and fruits with skin that is eaten. However, remember that washing can remove only a small proportion of most pesticides.

- Peeling does not remove all pesticides because many are stored in the flesh of fruits and vegetables. Peeling may do little more than waste valuable nutrients and dietary fibre.

- Buying fruit and vegetables from supermarkets carries the assurance of low residues and only a small chance of illegal excess residues going undetected.

- Organic crops should contain the minimum of residues, but try to buy from an approved grower, supermarket, or other trader you trust. Organic food is more expensive and not as widely available as other farm produce.

- If you have household timbers treated for dry rot or woodworm, ask the contractor what chemicals are being applied and check the safety precautions that *both* the contractor and you should follow.

- Ask your local authority to erect notices when parks or footpaths are being sprayed with fungicide or herbicide.

- If you live near a railway track (British Rail or underground) or a road, ask the controlling authority to notify you, *in advance*, of both the timing of routine sprayings and the chemicals used.

- If you live on or near farmland, or walk in the countryside, try asking local farmers when they apply chemicals during each cropping year and what they apply.

- Be vigilant and vocal. If you even suspect pesticides have been used carelessly or illegally, don't hesitate to report your suspicions to your local Environmental Health Officers in the Local Authority.

- If you are concerned about pesticides but feeling powerless, write to your MP and MEP asking what they are doing about pesticide safety. It is amazing but still true that when MPs get one letter a week on an issue they take notice; when they get two they start reading newspapers and checking with government departments; but when they get from three to five, they know there is a public issue on their hands, and they get moving for fear of missing out on the publicity. Haven't you got three or four friends who can write?

- There are many skilled and effective campaigning groups who not only know how to press for change in the public interest, but also need your support as voters and subscribers.

Friends of the Earth is the most influential pesticide campaigning group. It has a London head office and many regional groups

(26–28 Underwood Street, London N1 7JU). **Parents for Safe Food** campaigns for additive- and pesticide-reduced food, especially with the health of children in mind (Britannia House, 1–11 Glenthorne Road, London W6). The **Henry Doubleday Research Association** is interested in organic gardening (HDRA, National Centre for Organic Gardening, Ryton-on-Dunsmore, Coventry CV8 3LG). **The Soil Association** is also interested in organic farming and is one of several organisations which offer approval schemes to *bona fide* farmers who meet strict standards (86 Colston Street, Bristol BS1 5BB). Other organic grower organisations are the **Organic Growers' Association** and the **British Organic Growers**, both of which are at the Soil Association address. The **Pesticides Trust** is a valuable independent research organisation which compiles all manner of important information on pesticides in the environment (20 Compton Terrace, London N1).

PESTICIDE RESOURCE KIT

Introduction

The best decisions about pesticides require the best information. Some essential information is still withheld from public scrutiny, but much is available in hefty tomes and trade leaflets which are not often held in local libraries and are rarely sold in the high street. This resource kit was compiled with two aims: first, to support the discussion in the text; and, second, to help readers who have a special interest in, or concern with, pesticides to expand their knowledge and understanding.

1 **Crops and Their UK-Approved Pesticides** (pages 181–203).
   Active ingredients most likely to be used on crops, fields, and forests in the UK.

2 **Active Ingredients in Pesticides** (pages 204–85).
   Lists the major active ingredients approved for use in the UK, giving a description of their chemical type, uses and main precautions in use, together with a listing of the common products containing each active ingredient for industrial or general consumer use.

3 **Symptoms of Pesticide Poisoning** (pages 286–94).
   Any suspected pesticide poisoning should be taken to a doctor or hospital. Not all poisonings are fatal or even produce serious effects. The symptoms are often minor and can be difficult to distinguish from common ailments like rashes, breathing problems, and nausea. The table lists the symptoms of poisoning by the main types of pesticide and suggests first-aid measures.

4   **Appendices A and B:** Pesticide families and their modes of action   (pages 295–313).
This is an expanded version of the discussion in Chapter 1 and contains more chemical and other technical information than most readers will require.

Crops and Their UK-Approved Pesticides

Pesticide approvals in the UK limit the use of any pesticide to nominated crops. A crop grown on a British farm may not necessarily use all or any of the pesticides approved for use on that crop, but it is reasonable to expect that some of the allowed chemicals will be used unless it is an organic farm. The same restrictions do not apply to imported foods. Foreign exporters of, for example, tomatoes or flowers can use pesticides on their crops which UK farmers are not allowed to use. This, combined with the low level of residue testing, makes it difficult to choose crops which will be free of particular pesticides.

This table lists the main chemicals likely to be found on some major UK-grown crops. It has been compiled on the basis of the approved crop use which is specified on product labels. A manufacturer need not, of course, list all the crops for which the pesticide has been approved but only those for which there is a viable market. Some labels will state, for example, that the chemical is 'for use on arable crops', and we have listed these pesticides under a General Usage category. Other products will be labelled for use on specific crops e.g. sugar beet. On using the table, therefore, specific crop entries should be read in conjunction with any relevant general usage entry. These are cross referenced: Sugar Beet (1), for example, refers back to the General Usage category (1) Arable Crops.

The table is therefore a guide to the range and specific types of chemicals likely to be used on the main agricultural crops. Use of a pesticide does not mean that there will be residues in the food when bought. This is particularly so for chemicals like herbicides which may be applied to soil before the crop is planted. But knowing which chemicals can be sprayed on grass or fruit trees may give some idea as to which pesticides may be encountered during a stroll in the countryside. Use the table as a guide to the possible, and not as a list of what pesticides *are* used.

Foodstuffs

GENERAL USAGE – ARABLE CROPS
Category 1

Disease control   chloropicrin, dazomet, metham-sodium, methyl bromide

Pest control   aluminium ammonium sulphate, metaldehyde, methiocarb, tetraborate, ziram

Weed control   alphachloralose, amitrole, atrazine, azamethiphos, bioallethrin, bromadiolone, calciferol, chlorophacinone, coumatetralyl, difenacoum, fenitrothion, glyphosate, methodyl, paraquat, permethrin, phenothrin, propoxur, propyzamide, resmethrin, tetramethrin, warfarin

GENERAL USAGE – FIELD VEGETABLES
Category 2

Disease control   dazomet, etridiazole, propamocarb hydrochloride, thiabendazole, thiram

Pest control   aluminium ammonium sulphate, ammonium sulphamate, bromophos, captan, chlorpyrifos, lindane, metaldehyde, nicotine, pyrethrins, resmethrin, rotenone, trichlorfon

Weed control   ammonium sulphamate, dazomet, diquat, paraquat

GENERAL USAGE – FRUIT AND HOPS
Category 3

Pest control   aluminium ammonium sulphate, metaldehyde, nicotine, pyrethrins, resmethrin, rotenone, sodium tetraborate

Weed control   diquat, paraquat

GENERAL USAGE – PROTECTED CROPS
(under glass, under polythene tunnels, etc.)
Category 4

Disease control   chloropicrin, chlorothalonil, cresylic acid, dazomet, formaldehyde, fosetyl-aluminium, imazalil, metham-sodium, methyl bromide

Pest control   aldicarb, alphachloralose, chloropicrin, cresylic acid, cypermethrin, dazomet, deltamethrin, demeton-S-methyl, dichlorvos, dicofol, diflubenzuron, dimethoate, fenbutatin oxide, heptenophos, lindane, malathion, metaldehyde, methiocarb, nicotine, oxydemeton-methyl, pirimiphos-methyl, pyrethrins, rotenone, tetradifon

Weed control   benzalkonium chloride, chloropicrin, dazomet, dichlorophen, dodecylbenzyl trimethyl ammonium chloride, metham-sodium, methyl bromide

APPLES (3)

Disease control   benomyl, Bordeaux mixture, bupirimate, captan, carbendazim, copper oxychloride, cufraneb, fosetyl-aluminium, mancozeb, mercuric oxide, metalaxyl, myclobutanil, nitrothal-isopropyl, nuarimol, octhilinone, penconazole, pyrazophos, thiophanate-methyl, triadimefon, triforine

Pest control   amitraz, azinphos-methyl, carbaryl, chlorpyrifos, cyfluthrin, cypermethrin, deltamethrin, demeton-S-methyl sulphone, dicofol, diflubenzuron, dimethoate, fenitrothion, fenvalerate, heptenophos, lindane, malathion, nicotine, oxydemeton-methyl, pheromones, phosalone, pirimicarb, pirimiphos-methyl, pyrethrins, rotenone, tetradifon

Weed control   amitrole, asulam, dalaphon, 2,4-DES, dicamba, dichlobenil, diuron, glyphosate, isoxaben, MCPA, mecoprop, napropamide, oxadiazon, pendimethalin, pentanochlor, propyzamide, simazine, sodium monochloroacetate, terbacil, triclopyr

APRICOTS (3)

Disease control   Bordeaux mixture, copper hydroxide, copper oxychloride, copper sulphate, cufraneb, cupric ammonium carbonate, vinclozolin

Pest control   demeton-S-methyl, dimethoate, malathion, oxydemeton-methyl, tar oils, tetradifon

Weed control   amitrole, asulam

ASPARAGUS (2)

Disease control   iprodione, thiabendazole

Asparagus contd

Pest control	cypermethrin, heptenophos, nicotine, triazophos
Weed control	2,4-DES, diquat, diuron, MCPA, simazine, terbacil

AUBERGINES (4)

Disease control   iprodione, vinclozolin

Pest control   cypermethrin, fatty acids, heptenophos, nicotine, oxamyl, permethrin, resmethrin, vinclozolin

BEANS, FIELD (e.g. navy), BROAD (1)

Disease control   benomyl, captan, chlorothalonil, drazoxolon, fenpropimorph, iprodione, metalaxyl, methyl, thiabendazole, thiram

Pest control   aldicarb, bromophos, cyfluthrin, cypermethrin, deltamethrin, demeton-S-methyl, dimethoate, fatty acids, heptenophos, lindane, malathion, nicotine, pirimicarb, pyrethrins, resmethrin, triazophos

Weed control   alloxydim-sodium, bentazone, carbetamide, chlormequat, chlorpropham, diclofop-methyl, diuron, fenuron, glyphosate, prometryn, propyzamide, simazine, tri-allate, trifluralin

Crop control   chlormequat, di-l-p-menthene, diquat

BEANS, FRENCH/RUNNER (2)

Disease control   benomyl, carbendazim, drazoxolon, iprodione, metalaxyl, thiabendazole, thiophanate-methyl, vinclozolin

Pest control   *Bacillus thuringiensis*, bendiocarb, bromophos, cyfluthrin, cypermethrin, demeton-S-methyl, dimethoate, disulfoton, heptenophos, malathion, nicotine, oxydemeton-methyl, pirimicarb, tetradifon

Weed control   bentazone, chlorpropham, chlorthal-dimethyl, diclofop-methyl, diphenamid, diquat, monolinuron, trifluralin

BEETROOT (2)

Pest control   malathion, oxamyl, pyrazophos, pyrethrins, trichlorfon

Weed control   chlorpropham, clopyralid, ethofumesate, fenuron, metamitron, phenmedipham, propham, quizalofop-ethyl

BLUEBERRIES (3)
Weed control   asulam

BRASSICAS (2) (e.g. Brussels sprouts, cabbage, cauliflower, kale, kohlrabi, turnip)

Disease control benomyl, bromophos, captan, chlorothalonil, copper oxychloride, dichlofluanid, ferbam, fosetyl-aluminium, lindane, mancozeb, mercurous chloride, quintozene, thiophanate-methyl, thiram, tolclofos-methyl

Pest control   aldicarb, aluminium ammonium sulphate, azinphos-methyl, *Bacillus thuringiensis*, bromophos, carbofuran, chlorfenvinphos, chlorpyrifos, cyfluthrin, cypermethrin, demeton-S-methyl sulphone, diazinon, dimethoate, disulfoton, fenvalerate, fonofos, heptenophos, lindane, malathion, nicotine, permethrin, phorate, pirimicarb, pyrethrins, quinalphos

Weed control   aziprotryne, carbetamide, chlorthal-dimethyl, clopyralid, cyanazine, desmetryn, diclofop-methyl, metazachlor, propachlor, propyzamide, TCA, tebutam, tri-allate, trifluralin

BRASSICAS, ROOT (2) (e.g. mangolds)

Disease control benomyl, carbendazim, copper sulphate, dinocap, lindane, sodium propionate, thiabendazole, thiram, triadimefon, triadimenol, tridemorph

Pest control   aldicarb, carbofuran, carbosulfan, chlorfenvinphos, chlorpyrifos, cypermethrin, deltamethrin, dimethoate, fonofos, lindane, malathion, nicotine, pirimicarb, pyrazophos, trichlorfon

Weed control   alloxydim-sodium, dalapon, di-l-p-menthene, fluazifop-P-butyl, sethoxydim, TCA

BRASSICAS, SEED (1) (e.g. mustard)
Disease control iprodione

Brassicas, seed contd

Pest control   azinphos-methyl, endosulfan, malathion, phosalone

Weed control   benazolin, carbetamide, chlorthal-dimethyl,
clopyralid, diclofop-methyl, diquat, glyphosate,
propachlor, propyzamide, quizalofop-ethyl, tri-allate,
trifluralin

BRASSICAS, GROWN FOR SEED TO PLANT (2)

Disease control   iprodione

Pest control   lindane, phosalone

CABBAGES, IN STORAGE

Disease control   carbendazim, iprodione, metalaxyl

CANE FRUIT (3) (e.g. raspberries, loganberries)

Disease control   benodanil, benomyl, bupirimate, captan,
carbendazim, chlorothalonil, copper hydroxide,
copper oxychloride, copper sulphate, cufraneb,
cupric ammonium carbonate, dichlofluanid,
fenarimol, iprodione, mancozeb, metalaxyl, thiram,
triadimefon, vinclozolin

Pest control   azinphos-methyl, *Bacillus thuringiensis*, carbaryl,
chlorpyrifos, cresylic acid, cypermethrin,
deltamethrin, demeton-S-methyl sulphone, 1,3-
dichloropropene, dimethoate, fenitrothion,
heptenophos, lindane, malathion, oxydemeton-
methyl, pirimicarb, pyrethrins, rotenone, tar oils,
tetradifon

Weed control   asulam, atrazine, bromacil, chlorthal-dimethyl,
cyanazine, dalapon, 2,4-DES, dichlobenil, fluazifop-
P-butyl, isoxaben, lenacil, napropamide, paraquat,
pendimethalin, propyzamide, sethoxydim, simazine,
trifluralin

CARROTS (2)

Disease control   benomyl, copper hydroxide, iprodione, mancozeb,
metalaxyl, thiabendazole, triadimefon

Pest control   aldicarb, carbofuran, carbosulfan, demeton-S-methyl,
dimethoate, disulfoton, malathion, nicotine, phorate,
pirimicarb, pyrethrins, quinalphos, triazophos

Weed control	alloxydim-sodium, chlorpropham, diclofop-methyl, fenuron, fluazifop-P-butyl, linuron, metoxuron, pentanochlor, prometryn, TCA, tri-allate, trifluralin

CELERIAC (2)

Disease control	chlorothalonil
Pest control	chlorothalonil, chlorpropham, demeton-S-methyl, nicotine, pentanochlor, quinalphos, trifluralin
Weed control	chlorpropham, pentanochlor, trifluralin

CELERY (2)

Disease control	benomyl, Bordeaux mixture, bromophos, captan, carbendazim, chlorfenvinphos, chlorothalonil, cypermethrin, demeton-S-methyl, diazinon, dimethoate, disulfoton, heptenophos, malathion, nicotine, oxydemeton-methyl, phorate, pirimicarb, propamocarb hydrochloride, pyrethrins, quinalphos, trichlorfon, vinclozolin, zineb
Weed control	chlorpropham, diclofop-methyl, gibberellins, linuron, pentanochlor, prometryn

CEREALS (1)

Disease control	benodanil, benomyl, carbendazim, carboxin, chlorothalonil, copper hydroxide, copper oxychloride, ethirimol, ethylenebisdithiocarbamate, fenpropidin, ferbam, flusilazole, guazatine, imazalil, iprodione, lindane, mancozeb, maneb, 2-methoxyethylmercury acetate, oxycarboxin, phenylmercury acetate, potassium sorbate, prochloraz, propiconazole, pyrazophos, sodium metabisulphite, sulphur, thiabendazole, thiophanate-methyl, thiram, tridemorph, zineb
Pest control	alphacypermethrin, aluminium ammonium sulphate, bifenthrin, chlorfenvinphos, chlorpyrifos, cyfluthrin, cypermethrin, deltamethrin, demeton-S-methyl, dimethoate, fenitrothion, fenvalerate, fonofos, heptenophos, omethoate, oxydemeton-methyl, phosalone, pirimicarb, quinalphos
Weed control	amitrole, benazolin, bifenox, bromoxynil, chlorotoluron, chlorsulfuron, clopyralid, cyanazine, 2,4-D, 2,4-DB, dalapon, dicamba, diclofop-methyl,

Cereals contd

fluroxypyr, glyphosate, imazamethabenz-methyl, ioxynil, isoproturon, isoxaben, linuron, MCPB, paraquat, pendimethalin, pyridate, TCA, terbutryn, tri-allate, trifluralin

Crop control   chlormequat, 2-chloroethylphosphonic acid, chlorpropham, choline chloride, di-l-p-menthene, paclobutrazol

CEREALS AND RAPESEED, IN STORAGE

Pest control   chloropicrin, chlorpyrifos-methyl, etrimfos, fenitrothion, methyl bromide, permethrin, pirimiphos-methyl, resmethrin

CHERRIES (4)

Disease control   benodanil, Bordeaux mixture, carbendazim, copper hydroxide, copper oxychloride, copper sulphate, cufraneb, octhilinone, tar oils, vinclozolin

Pest control   azinphos-methyl, chlorpyrifos, clofentezine, cyfluthrin, cypermethrin, deltamethrin, demeton-S-methyl, demeton-S-methyl sulphone, fenitrothion, malathion, omethoate, oxydemeton-methyl, pheromones, pirimicarb

Weed control   amitrole, asulam, 2,4-DES, glyphosate, isoxaben, napropamide, pendimethalin, pentanochlor, propyzamide, simazine, sodium monochloroacetate

CHICORY (2)

Disease control   quintozene

Pest control   cypermethrin, heptenophos, nicotine

Weed control   propyzamide

CRANBERRIES *see* BLUEBERRIES

CUCUMBERS, PROTECTED (4)

Disease contrd   benomyl, bupirimate, carbendazim, chlorothalonil, copper oxychloride, cupric ammonium carbonate, dicloran, dinocap, etridiazole, fenarimol, imazalil, iprodione, metalaxyl, nitrothal-isopropyl, propamocarb hydrochloride, pyrazophos, quintozene, thiram, triforine, vinclozolin, zineb-ethylene thiuram disulphide complex

Pest control   *Bacillus thuringiensis*, cypermethrin, deltamethrin, demeton-S-methyl, diazinon, dicofol, fatty acids, fenbutatin oxide, heptenophos, permethrin, pirimicarb, pirimiphos-methyl, propoxur, pyrazophos, pyrethrins, resmethrin

CUCURBITS, OUTDOOR (2) (e.g. cucumbers, courgettes, marrows)

Disease control   bupirimate, imazalil, iprodione, maneb propamocarb hydrochloride, thiabendazole, thiram, zineb

Pest control   bendiocarb, cypermethrin, demeton-S-methyl, dimethoate, diphenamid, disulfoton, heptenophos, nicotine, pirimicarb, pyrazophos

Weed control   diphenamid

CUCURBITS, PROTECTED (GLASSHOUSE) (4)

Disease control   propamocarb hydrochloride

Pest control   bendiocarb, cypermethrin, fatty acids, pirimicarb, nicotine, tetradifon

CURRANTS (3) (e.g. blackcurrants, redcurrants)

Disease control   benomyl, Bordeaux mixture, bupirimate, carbendazim, chlorothalonil, copper hydroxide, cufraneb, dichlofluanid, dinocap, mancozeb, maneb, triforine, vinclozolin, zineb

Pest control   azinophos-methyl, carbaryl, chlorpyrifos, cresylic acid, cyfluthrin, cypermethrin, demeton-S-methyl, dicofol, diflubenzuron, dimethoate, endosulfan, fenitrothion, heptenophos, malathion, methiocarb, oxydemeton-methyl, pirimicarb, pyrethrins

Weed control   asulam, chlorpropham, chlorthal-dimethyl, dalapon, 2,4-DES, dichlobenil, diuron, fluazifop-P-butyl, isoxaben, lenacil, MCPB, napropamide, oxadiazon, paraquat, pendimethalin, pentanochlor, propyzamide, simazine, sodium monochloroacetate

DAMSONS *see* CHERRIES

FENNEL (2)

Pest control   cypermethrin, deltamethrin, nicotine, pirimicarb, pyrazophos, quinalphos, triazophos

Weed control   chlorpropham, pentanochlor

FRUIT TREES, NURSERY STOCK (3)

Disease control   carbendazim, dodemorph, iprodione

Pest control   aldicarb, amitraz, asulam, azinphos-methyl, clopyralid, deltamethrin, demeton-S-methyl sulphone, fonofos, linuron, napropamide, propyzamide, triazophos, trietazone, trifluralin

FRUIT TREE (3)

Disease control   octhilinone, *Trichoderma viride*

Pest control   aluminium ammonium sulphate, cresylic acid, fatty acids, malathion, pheromones, pyrethrins, rotenone, ziram

Weed control   dicamba, glyphosate, MCPA, mecoprop, paraquat

GARLIC (4)

Pest control   deltamethrin, oxamyl

GOOSEBERRIES (3)

Disease control   benomyl, Bordeaux mixture, bupirimate, carbendazim, chlorothalonil, copper hydroxide, copper sulphate, cufraneb, dichlofluanid, dodine, fenarimol, mancozeb, triadimefon, vinclozolin

Pest control   azinphos-methyl, carbaryl, chlorpyrifos, cypermethrin, dimethoate, fenitrothion, heptenophos, malathion, nicotine, oxydemeton-methyl, pirimicarb, pyrethrins, rotenone, tar oils

Weed control   asulam, chlorpropham, chlorthal-dimethyl, dalapon, 2,4-DES, diuron, fluazifop-P-butyl, isoxaben, lenacil, napropamide, oxadiazon, paraquat, pendimethalin, pentanochlor, propyzamide, simazine, sodium monochloroacetate

GRAPEVINES (3)

Disease control Bordeaux mixture, chlorothalonil, copper hydroxide, copper oxychloride, cufraneb, dichlofluanid, fenarimol, mancozeb, metalaxyl, triadimefon, vinclozolin

Pest control cypermethrin, dimethoate, petroleum oil, tetradifon

Weed control diquat, glyphosate, oxadiazon, paraquat, propyzamide, simazine

HERBS (2)

Pest control deltamethrin, demeton-S-methyl

Weed control chlorthal-dimethyl, clopyralid, diquat, lenacil, linuron, propachlor, propyzamide, trifluralin

HOPS (3)

Disease control benomyl, Bordeaux mixture, bupirimate, carbendazim, chlorothalonil, copper hydroxide, dinocap, fosetyl-aluminium, metalaxyl, penconazole, pyrazophos, triadimefon, triforine, zineb, zineb-ethylene thiuram disulphide complex

Pest control aldicarb, amitraz, cypermethrin, deltamethrin, demeton-S-methyl, 1,3-dichloropropene, dicofol, dimethoate, endosulfan, fenvalerate, lindane, malathion, mephosfolan, nicotine, omethoate, pyrethrins

Weed control asulam, diquat, oxadiazon, paraquat, pendimethalin, propyzamide, simazine, sodium monochloroacetate

LEEKS (2)

Disease control benomyl, bromophos, carbendazim, chlorothalonil, copper hydroxide, fenpropimorph, ferbam, iodofenphos, iprodione, mancozeb, maneb, mercurous chloride, metalaxyl, metham-sodium, propamocarb hydrochloride, thiabendazole, triadimefon, zineb

Pest control aldicarb, benomyl, bromophos, carbofuran, cypermethrin, deltamethrin, demeton-S-methyl, diazinon, dimethoate, fenitrothion, heptenophos, iodofenphos, malathion, metalaxyl, metham-sodium, nicotine, oxamyl, pyrethrins, thiabendazole, triazophos

Leeks contd

Weed control alloxydim-sodium, aziprotryne, bentazone, chlorbufam, chloridazon, chlorpropham, chlorthal-dimethyl, clopyralid, cyanazine, diclofop-methyl, fenuron, fluazifop-P-butyl, ioxynil, monolinuron, prometryn, propachlor, tri-allate

LEGUMES (1) (e.g. vetch and field peas for animal feed)

Disease control thiabendazole, thiram

Pest control demeton-S-methyl

Weed control asulam, benazolin, carbetamide, chlorpropham, clopyralid, dalapon, 2,4-DB, diclofop-methyl, fluazifop-P-butyl, paraquat, propyzamide, tri-allate

Crop control diquat

LETTUCE, OUTDOOR (2)

Disease control benomyl, carbendazim, iprodione, mancozeb, metalaxyl, quintozene, thiram, tolclofos-methyl, vinclozolin, zineb

Pest control carbaryl, cypermethrin, demeton-S-methyl, diazinon, dimethoate, heptenophos, lindane, malathion, nicotine, phorate, pirimicarb, pyrethrins, resmethrin

Weed control chlorpropham, diclofop-methyl, diuron, propachlor, propyzamide, trifluralin

LETTUCE, PROTECTED (4)

Disease control dicloran, fosetyl-aluminium, iprodione, lindane, metalaxyl, tecnazene, thiram, tolclofos-methyl, vinclozolin

Pest control lindane, malathion, oxydemeton-methyl, pirimicarb, tecnazene

Weed control chlorpropham, propyzamide

LINSEED (1)

Disease control captan, iprodione, lindane, thiabendazole, thiram

Weed control bentazone, bromoxynil, clopyralid, diclofop-methyl, diquat, glyphosate, linuron, tri-allate, trifluralin

Crop control diquat

MANGE-TOUT (1)
Disease control   drazoxolon, iprodione, metalaxyl, thiabendazole, thiram
Pest control   demeton-S-methyl, nicotine, triazophos
Weed control   MCPB, simazine, tri-allate, trietazine

MUSHROOMS (4)
Disease control   benomyl, carbendazim, chlorothalonil, dichlorophen, prochloraz, sodium pentachlorophenoxide, thiabendazole, zineb
Pest control   chlorfenvinphos, deltamethrin, diazinon, dichlorvos, diflubenzuron, malathion, nicotine, permethrin, pirimiphos-ethyl, pirimiphos-methyl, pyrethrins, resmethrin
Weed control   sodium pentachlorophenoxide

NECTARINES *see* APRICOTS

NUTS (3) (tree nuts e.g. Kent cobs, not peanuts)
Disease control   benomyl, iprodione
Pest control   oxydemeton-methyl
Weed control   glyphosate

OILSEED RAPE (1) (*see also* Cereals and Rapeseed, in storage)
Disease control   benomyl, carbendazim, chlorothalonil, ferbam, iprodione, lindane, mancozeb, maneb, potassium, prochloraz, propiconazole, sodium metabisulphite, sorbate, sulphur, thiram, vinclozolin, zineb
Pest control   aldicarb, alphacypermethrin, aluminium ammonium sulphate, azinphos-methyl, bifenthrin, carbaryl, carbofuran, carboxin, chlorpyrifos, cyfluthrin, deltamethrin, dimethoate, endosulfan, fenvalerate, fonofos, lindane, malathion, phorate, thiabendazole
Weed control   alachlor, benazolin, carbetamide, clopyralid, cyanazine, dalapon, diclofop-methyl, glyphosate, metazachlor, napropamide, propachlor, propyzamide, pyridate, quizalofop-ethyl, sethoxydim, TCA, tebutam, trifluralin
Crop control   chlormequat, di-l-p-menthene, diquat

ONIONS, OUTDOOR *see* **LEEKS**
ONIONS, PROTECTED *see* **GARLIC**
PARSLEY *see* **CARROTS**
PARSNIPS *see* **CARROTS**
PEACHES *see* **APRICOTS**
PEARS *see* **APPLES**

PEAS (1)

Disease control   benomyl, captan, carbendazim, chlorothalonil, drazoxolon, iprodione, thiabendazole, thiram, vinclozolin

Pest control   aldicarb, aluminium ammonium sulphate, azinphos-methyl, bifenthrin, carbaryl, cyfluthrin, cypermethrin, deltamethrin, dimethoate, fenitrothion, fenvalerate, lindane, oxamyl, triazophos

Weed control   alloxydim-sodium, aziprotryne, bentazone, chlorpropham, cyanazine, diclofop-methyl, diuron, fenuron, glyphosate, MCPA, MCPB, prometryn, simazine, TCA, terbuthylazine, terbutryn, tri-allate, trifluralin

Crop control   chlormequat, di-l-p-menthene, diquat

PEPPERS (4)

Disease control   carbendazim, chlorothalonil, etridiazole, iprodione, propamocarb hydrochloride, vinclozolin

Pest control   *Bacillus thuringiensis*, cypermethrin, diflubenzuron, etridiazole, nicotine, oxamyl, tetradifon

POTATOES (1)

Disease control   benalaxyl, Bordeaux mixture, chlorothalonil, copper sulphate, cufraneb, cymoxanil, fentin acetate, fentin hydroxide, ferbam, imazalil, iprodione, mancozeb, maneb, metoxuron, pencycuron, thiabendazole, tolclofos-methyl

Pest control   aldicarb, alphacypermethrin, azinphos-methyl, carbaryl, chlorfenvinphos, chlorpyrifos, cypermethrin, deltamethrin, disulfoton, endosulfan, heptenophos, malathion, nicotine, phorate, pirimicarb, pyrethrins, thiofanox, triazophos

Weed control	alloxydim-sodium, bentazone, cyanazine, dalapon, diclofop-methyl, diquat, EPTC, linuron, metribuzin, monolinuron, pendimethalin, prometryn, TCA, terbutryn

POTATOES, IN STORAGE

Disease control	carbendazim, chlorpropham, ethanol-iodine complex, maleic hydrazide, nonylphenoxypoly (ethyleneoxy) ethanol-iodine complex, propham, tecnazene, thiabendazole

PLUMS *see* CHERRIES

QUINCES (3)

Weed control	amitrole

RADICCHIO (2)

Disease control	iprodione
Pest control	deltamethrin, demeton-S-methyl
Weed control	propyzamide

RHUBARB (2)

Pest control	chlorpyrifos, cypermethrin, heptenophos, nicotine, triazophos
Weed control	chlorpropham, dalapon, 2,4-DES, fenuron, propyzamide, TCA

SORREL (2)

Pest control	deltamethrin

SPINACH (2)

Disease control	propamocarb hydrochloride, zineb
Pest control	cypermethrin, demeton-S-methyl, heptenophos, nicotine, thiram, trichlorfon
Weed control	chlorpropham, fenuron, lenacil

STRAWBERRIES

Disease control   benomyl, bupirimate, captan, carbendazim, chlorothalonil, copper oxychloride, dichlofluanid, dinocap, etridiazole, fenarimol, fosetyl-aluminium, iprodione, propamocarb hydrochloride, thiram, triadimefon, vinclozolin

Pest control   aldicarb, carbaryl, chlorpyrifos, cyfluthrin, cypermethrin, demeton-S-methyl, 1,3-dichloropropene, dicofol, dimethoate, disulfoton, endosulfan, fenbutatin oxide, heptenophos, lindane, malathion, methiocarb, nicotine, oxydemeton-methyl, phorate, pirimicarb, pyrethrins

Weed control   alloxydim-sodium, asulam, chloroxuron, chlorpropham, chlorthal-dimethyl, clopyralid, diphenamid, diquat, ethofumesate, fenuron, fluazifop-P-butyl, lenacil, napropamide, pendimethalin, phenmedipham, propachlor, propyzamide, sethoxydim, simazine, terbacil, trifluralin

SUGARBEET (1)

Disease control   fentin hydroxide, hymexazol, maneb, thiram, triadimefon, triadimenol

Pest control   aldicarb, alphacypermethrin, benfuracarb, carbofuran, carbosulfan, chlorpyrifos, deltamethrin, dimethoate, disulfoton, heptenophos, lindane, nicotine, phorate, pirimicarb, pirimiphos-methyl, thiofanox

Weed control   alloxydim-sodium, carbetamide, chloradizan, clopyralid, dalapon, diclofop-methyl, ethofumesate, fenuron, glyphosate, lenacil, metamitron, paraquat, phenmedipham, propham, TCA, tri-allate, trifluralin

SUNFLOWERS (1)

Weed control   diquat, trifluralin

SWEETCORN (1)

Disease control   thiram

Pest control   aldicarb, bendiocarb, carbofuran, chlorfenvinphos, chlorpyrifos, demeton-S-methyl, dimethoate, fenitrothion, phorate, triazophos

Weed control   alachlor, atrazine, clopyralid, cyanazine, 2,4-DES, difenzoquat, pyridate, simazine, tri-allate

TOMATOES, OUTDOOR (2)

Disease control   copper oxychloride, metham-sodium

Pest control   metham-sodium

TOMATOES, PROTECTED (4)

Disease control   benomyl, bupirimate, captan, carbendazim, chlorothalonil, copper hydroxide, copper oxychloride, copper sulphate, cufraneb, cupric ammonium carbonate, dichlofluanid, dicloran, iprodione, lindane, maneb, metham-sodium, nabam, nonylphenoxypoly (ethyleneoxy)ethanol-iodine complex, propamocarb hydrochloride, quintozene, tecnazene, thiram, vinclozolin, zinc, zineb

Pest control   aldicarb, *Bacillus thuringiensis*, carbaryl, cypermethrin, deltamethrin, demeton-S-methyl, diazinon, diflubenzuron, dimethoate, fenbutatin oxide, lindane, malathion, nicotine, oxamyl, permethrin, pirimicarb, propoxur, pyrethrins, resmethrin, tecnazene, trichlorfon

Weed control   diphenamid, metham-sodium, pentanochlor

Plant growth regulation   2-chloroethylphosphonic acid, (2-naphthyloxy) acetic acid

WATERCRESS (2)

Disease control   benomyl, mancozeb, metalaxyl

Pest control   dimethoate, malathion

Flowers and Ornamental Plants

GENERAL USAGE

Disease control   bupirimate, captan, carbendazim, cupric ammonium carbonate, fenarimol, 8-hydroxyquinoline sulphate, imazalil, metham-sodium, propamocarb hydrochloride, tecnazene, thiabendazole, tolclofos-methyl, vinclozolin

General usage contd

Pest control	aldicarb, aluminium ammonium sulphate, *Bacillus thuringiensis*, carbaryl, clofentezine, deltamethrin, demeton-S-methyl, diazinon, diflubenzuron, dimethoate, fenbutatin oxide, fonofos, heptenophos, lindane, malathion, metaldehyde, metham-sodium, nicotine, oxamyl, permethrin, pirimicarb, pyrazophos, pyrethrins, resmethrin, rotenone, sodium tetraborate, triazophos
Weed control	alloxydim-sodium, ammonium sulphamate, chlorthal-dimethyl, diquat, linuron, metazachlor, paraquat, trifluralin *for use only on standing ground, not on plants* chlospropham, fenuron
Plant growth regulation	dikegulac, 4-indol-3-ylbutyric acid, 1-naphthylacetic acid

ANNUALS/BIENNIALS

Weed control	chlorpropham, fenuron, pentanochlor

BEDDING PLANTS

Disease control	benomyl, carbendazim, furalaxyl, iprodione, quintozene
Pest control	diazinon, endosulfan, pirimiphos-ethyl
Weed control	chloramben
Plant growth regulation	chlormequat, daminozide, paclobutrazol

BULBS/CORMS

Disease control	benomyl, carbendazim, chlorothalonil, dichlofluanid, etridiazole, ferbam, iprodione, mancozeb, maneb, propamocarb hydrochloride, zineb, zineb-ethylene-thiuram disulphide complex
Pest control	aldicarb, aluminium ammonium sulphate, carbofuran, 1,3-dichloropropene, disulfoton, endosulfan, nicotine
Weed control	bentazone, chlorbufam, chloridazon, chlorpropham, cyanazine, 2,4-DES, diphenamid, diquat, diuron, fenuron, lenacil, linuron, paraquat, pentanochlor

Plant growth regulation   2-chloroethylphosphonic acid, paclobutrazol

CHRYSANTHEMUMS

Disease control   benodanil, benomyl, bupirimate, captan, carbendazim, chlorothalonil, cupric ammonium carbonate, dinocap, iprodione, lindane, mancozeb, metham-sodium, nabam, oxycarboxin, propiconazole, quintozene, tecnazene, thiram, triforine, zineb

Pest control   aldicarb, *Bacillus thuringiensis*, demeton-S-methyl, diazinon, lindane, nicotine, permethrin, propoxur, resmethrin, tecnazene

Weed control   chloroxuron, chlorpropham, metham-sodium, pentanochlor

Plant growth regulation   daminozide, paclobutrazol

CONTAINER-GROWN STOCK

Disease control   etridiazole, propamocarb hydrochloride

Pest control   diazinon

Weed control   carbetamide, chloramben, chlorothalonil, chloroxuron, napropamide, oryzalin, oxadiazon

DAHLIAS

Disease control   quintozene

Pest control   aldicarb

Weed control   lenacil

FLOWERS, CUT (GLASSHOUSE)
(for general use pesticides *see* Protected Crops page 182)

Disease control   benomyl, carbendazim, dicloran, dinocap, imazalil, iprodione, mancozeb, metham-sodium, oxycarboxin, quintozene, thiram, triforine

Pest control   aldicarb, demeton-S-methyl, dimethoate, endosulfan, malathion, pirimicarb, propoxur, resmethrin

Weed control   metham-sodium, pentanochlor

Plant growth regulation   2-chloroethylphosphonic acid, sodium silver thiosulphate

HARDY ORNAMENTAL NURSERY STOCK

Disease control   carbendazim, chlorothalonil, dodemorph, drazoxolon, etridiazole, fosetyl-aluminium, furalaxyl, iprodione, prochloraz, propamocarb hydrochloride

Pest control   aldicarb, amitraz, azinphos-methyl, carbofuran, deltamethrin, demeton-S-methyl sulphone, fonofos

Weed control   asulam, atrazine, chlorpropham, clopyralid, dalapon, 2,4-DES, diphenamid, diuron, glyphosate, lenacil, linuron, metazachlor, napropamide, oryzalin, paraquat, pentanochlor, propyzamide, simazine, trietazone, trifluralin

Plant growth regulation   paclobutrazol

HEDGES

Pest control   trichlorfon

Weed control   dalapon, dichlobenil, diquat, glyphosate, paraquat, simazine

Plant growth regulation   dikegulac, maleic hydrazide

ORNAMENTAL TREES AND SHRUBS

Disease control   cresylic acid, metham-sodium, octhilinone, penconazole, thiabendazole, *Trichoderma viride*

Pest control   aldicarb, deltamethrin, diflubenzuron, dimethoate, fatty acids, lindane, permethrin, ziram

Weed control   chloramben, clopyralid, cyanazine, dalapon, dichlobenil, diquat, diuron, glyphosate, napropamide, oxadiazon, propachlor, propyzamide

Plant growth regulation   dikegulac, maleic hydrazide, 1-naphthylacetic acid

PERENNIALS

Disease control   dichlofluanid

Weed control   chloramben, chloroxuron, chlorpropham, 2,4-DES, fenuron, propachlor, simazine

POT PLANTS
(for general use pesticides *see* Protected Crops page 182)

Disease control	benomyl, captan, carbendazim, furalaxyl, iprodione, lindane, mancozeb, oxycarboxin, pyrazophos, quintozene, tecnazene, thiram
Pest control	aldicarb, clofentezine, cypermethrin, diazinon, endosulfan, heptenophos, lindane, permethrin, pirimiphos-ethyl, propoxur, resmethrin
Weed control	chloroxuron, pentanochlor, trimethyl ammonium chloride
Plant growth regulation	chlormequat, 2-chloroethylphosphonic acid, daminozide, dikegulac, dodecylbenzyl, paclobutrazol, sodium silver thiosulphate, trimethylammonium chloride

ROSES

Disease control	benodanil, bupirimate, captan, carbendazim, chlorothalonil, dichlofluanid, dinocap, dodemorph, fenarimol, imazalil, mancozeb, maneb, myclobutanil, penconazole, pyrazophos, thiram, triforine
Pest control	dimethoate, malathion, nicotine, rotenone
Weed control	alloxydim-sodium, atrazine, dalapon, 2,4-DES, dichlobenil, lenacil, pentanochlor, propyzamide, simazine
Plant growth regulation	1-naphthylacetic acid, paclobutrazol

WATER LILY

Disease control	carbendazim, etridiazole, furalaxyl, metalaxyl

Grassland

LEY CROPS

Disease control	prochloraz
Pest control	cypermethrin, omethoate
Weed control	benazolin, bromoxynil, cyanazine, dicamba, ethofumesate, ioxynil, linuron, MCPB, mecoprop, methabenzthiazuron

SOWN OR NATURAL PASTURE

Disease control   chlorpyrifos, fonofos, lindane, pirimicarb,
                    propiconazole, triadimefon, triazophos

Weed control   asulam, benazolin, bentazone, bromoxynil,
                    clopyralid, cyanazine, 2,4-D, dicamba, ethofumesate,
                    fluroxypyr, linuron, MCPA, mecoprop

Turf/Amenity Grass

Disease control   benodanil, benomyl, carbaryl, carbendazim,
                    chlordane, chlorothalonil, chloroxuron, chlorpyrifos,
                    dalapon, dicamba, dichlorophen, dichlorprop,
                    ferrous sulphate, ioxynil, iprodione, lindane, maleic
                    hydrazide, MCPA, mecoprop, oxycarboxin,
                    paclobutrazol, picloram, quintozene, thiabendazole,
                    thiophanate-methyl, triclopyr, triforine, vinclozolin

Forestry

NURSERY BEDS

Pest control   aldicarb, lindane, pirimicarb
Weed control   diphenamid, paraquat, simazine

TRANSPLANT LINES

Pest control   chlorpyrifos, lindane
Weed control   paraquat, simazine

PLANTATIONS

Pest control   aluminium phosphide, diflubenzuron, lindane,
                    resmethrin, warfarin, ziram

Weed control   ammonium sulphamate, asulam, atrazine, clopyralid,
                    cyanazine, 2,4-D, dalapon, dicamba, dichlobenil,
                    diquat, glyphosate, hexazinone, paraquat,
                    propyzamide, simazine, terbuthylazine, triclopyr

FARM WOODLAND

Weed control   propyzamide

CUT LOGS/TIMBER

Pest control   chlorpyrifos, lindane

Total Vegetation Control

e.g. around footpaths, playgrounds, roadsides, and individual buildings, railway lines

Pest control   bromadiolone

Weed control   amitrole, asulam, atrazine, benzalkonium chloride, bromacil, chloroxuron, chlorpropham, clopyralid, copper sulphate, cresylic acid, 2,4-D, dalapon, dicamba, dichlobenil, dichlorophen, diquat, fenuron, fosamine-ammonium, imazapyr, MCPA, paraquat, pentachlorophenol, picloram, simazine, sodium chlorate, sodium tetraborate, tebuthiuron, triclopyr, trimethylammonium chloride

WEEDS IN OR NEAR WATER

Weed control   asulam, 2,4-D, dalapon, dichlobenil, diquat, fosamine-ammonium, glyphosate, terbutryn

Plant growth regulation   maleic hydrazide

Non-crop Pest Control

Disease control   formaldehyde, lindane

Pest control   alphachloralose, azamethiphos, bioallethrin, bromadiolone, calciferol, carbaryl, chlorophacinone, coumatetralyl, dichlorvos, difenacoum, fenitrothion, methomyl, permethrin, phenothrin, pyrethrins, resmethrin, tetramethrin, warfarin

Active Ingredients in Pesticides

Most pesticides contain several ingredients. There is usually more than one active ingredient; the chemicals which are responsible for the specific action, as well as other chemicals known as adjuvants. The adjuvants are responsible for the physical properties of the pesticide product. They include solvents, surfactants (which allow liquids to wet surfaces they contact), and extenders, which allow the small volumes of active ingredients to be diluted into a larger volume which can be handled more easily. The list of active ingredients is shown on the label of every pesticide product approved for sale in the UK.

Pesticide products are approved mixtures of active ingredients plus their adjuvants. There are over 400 approved products in the UK. Most of these are limited to professional use, on farms or in industry. However, many of the ingredients are common to products approved for household or garden use. Often the same product is approved for use in households *and* in industry.

The table is a guide to the ingredients lists found on products and to the chemical names found increasingly in books, newspapers, and television programmes. The main active ingredients approved for use in the UK are listed. Not every one of the 400-plus approved combinations of ingredients is listed separately. However, under each active ingredient heading, the other chemicals which have been approved for mixing with it are detailed. Particulars of each approved mixture can be found in the references below.

How to use the table

Function   Describes the pesticidal action of the chemical. The type or family of the chemical is included in square brackets if it is included in the list of major families described in Chapter 1.

Uses	Lists the main uses for which the chemical has approval.
Single ingredient products	Includes all products marketed with that ingredient as the sole active ingredient. (They may, however, also contain other supporting chemicals to enable the pesticide to be applied effectively.) Product names in **bold** are those which can be sold to the general public, rather than only for professional or industrial uses.
Permitted mixtures	Lists the other active ingredients which have been approved for use with this active ingredient in various significant products. The combinations of these ingredients are specific to each approved product. Little-used mixtures are not included. Further details of mixtures and products can be found in the official MAFF publication of approved chemicals (see below).
Products based on mixtures	Lists the main products based on mixtures, again with those available to the general public given in **bold**. All products listed contain the main active ingredient and at least one other from the list of permitted mixtures. Check the label against the MAFF publication for the specific ingredients in any product.
Risks and precautions	Covers some of the more important health and environmental risks from using the chemical, and lists precautions which are recommended to reduce risks. These are specified when chemicals are approved, and most are included on the labels or the data sheets which accompany products. For reasons of space, the table does not include all the risks and precautions, but it highlights those which are most significant. The manufacturer is obliged to provide consumers with a copy of the data sheet on request. Contact the address on the label.
Status	Many of the approved chemicals are undergoing review or have reviews planned as part of the government's programme of reviewing all those older chemicals whose original approvals were based on less rigorous testing than that followed today. Reviews can also be ordered at any time if new evidence throws doubt on the list of risks or on the justification for an approval. This information may change at any time but is a

useful guide to chemicals whose approvals may
be revoked or at least whose conditions of
approval could be restricted in the near future.
Where status is not specified, the active
ingredient is, as of January 1991, fully approved
and not subject to a review. The MAFF will be
able to confirm any changes in status.

Further details of approved products are available in: MAFF-HSE,
Pesticides 1990, London, HMSO, 1990; Ivens, G. W. (ed.), *The UK
Pesticide Guide*, Wallingford, Common Agricultural Bureau and British
Crop Protection Council, 1990; Worthing, Charles R. (ed.), *The Pesticide
Manual: A World Compendium* (8th edn), Thornton Heath, British Crop
Protection Council, 1987.

Alachlor

Function	Residual pre-emergence herbicide [amides and ureas].
Uses	Against annual dicotyledons and grasses in rape and maize.
Single ingredient products	Lasso.
Permitted mixtures	None significant.
Risks and precautions	Harmful if swallowed, irritates eyes and skin.
Status	Under review.

Aldicarb

Function	Soil-acting insecticide and nematicide [carbamate].
Uses	Wide application against aphids, cabbage root fly, leaf miners, mealy bugs, millipedes in crops like beets, cabbage, potatoes, tomatoes, strawberries, and glasshouse crops like tomatoes

and flowers; thrips, whitefly, and frit fly in sweet
corn. Used against nematodes in potatoes,
cabbage, beans, strawberries, and onions.

Single ingredient products	Temix 10G.
Permitted mixtures	With lindane.
Risks and precautions	Toxic through skin, breathing, or swallowing. Irritating to eyes, skin, lungs. Hazardous to wildlife. An anticholinesterase compound. Not to be used on pot plants within 4 weeks of sale.

Alkylaryl trimethylammonium chloride

Function	Moss and algae killer [heterocyclic].
Uses	Against mosses, algae, and liverworts in pot plants and paths, and as mould inhibitor in paints and surface treatments for wood and various walls.
Single ingredient products	Gloquat in agriculture, wide range of wood and surface treatments.
Permitted mixtures	None significant.
Risks and precautions	Dangerous to fish.

Alloxydim-sodium

Function	Systemic, post-emergence herbicide.
Uses	Against grass weeds in crops like beans, swedes, potatoes, carrots, onions, strawberries, ornamentals.
Single ingredient products	Clout, **Weed out**.
Permitted mixtures	None significant.
Risks and precautions	Irritates skin and eyes; harvest interval: 4 weeks on all crops except beets (8 weeks).

Alphachloralose

Function	Rodenticide.
Uses	Against mice in farm buildings and glasshouses.

Single ingredient products	Alphachloralose concentrate, Townex mouse poison.
Permitted mixtures	None significant.
Risks and precautions	Toxic if swallowed; keep away from children and domestic animals.
Status	Review planned.

Alphacypermethrin

Function	Contact and internal insecticide [pyrethroid] for crops.
Uses	Against aphids in cereals, cabbage stem flea beetle in rape, cutworms in potatoes and beets. Restricted use on flowering rape against weevils, aphids, and blossom beetles.
Single ingredient products	Fastac, Fendona.
Permitted mixtures	None significant.
Risks and precautions	Harmful if swallowed, contacts skin, or gets in eyes; irritates skin; dangerous to bees and fish.

2-aminobutane

Function	Antifungal fumigant.
Uses	To protect potatoes in storage.
Single ingredient products	CSC 2-aminobutane.
Permitted mixtures	None significant.
Risks and precautions	Harmful if breathed. Irritating to skin, lungs, and eyes.
Status	Under review.

Amitraz

| Function | Insecticide and acaricide for tree fruit and hops. |
| Uses | Against aphids, red spider mites, and codling moths in apples, pears, hops, and ornamentals or nursery fruit trees and bushes. |

Single ingredient products	Mitac 20.
Permitted mixtures	None significant.
Risks and precautions	Harmful if swallowed or contacts skin.

Amitrole

Function	General, systemic leaf-acting herbicide [heterocyclic, triazole].
Uses	Against annual and perennial weeds like coltsfoot, dock, horsetail, couch grass in non-crop areas and crops like cereals and fruit.
Single ingredient products	Weedazol TL, MSS Aminotriazole.
Permitted mixtures	With atrazine, bromacil, 2,4-D, dicamba, diquat, diuron, MCPA, paraquat, simazine.
Products based on mixtures	Atlas total, Boroflow, Sarapron, Bullseye, Groundhog, Meganox, Destral, Hytrol, Mascot Highway, **Boots Long-Lasting Weedkiller**, **Murphy Path Weed Killer**, **Hytrol**, **Pathclear**, **Murphy Problems Weedkiller**, **Super Weedex**, **Tesco's Total Weedkiller Liquid**.
Risks and precautions	Irritating to skin and eyes; harmful to fish.

Ammonium sulphamate

Function	General herbicide and tree-killer.
Uses	For general preplanting weed control in vegetable and ornamental crops, and in forestry.
Single ingredient products	**Amcide**, **Root out**.
Permitted mixtures	None significant.
Risks and precautions	Irritating to skin and eyes; harmful to fish.

Asulam

Function	Systemic herbicide [carbamate].
Uses	For control of dock and bracken in pasture, uplands, tree and bush fruit crops, nursery crops.
Single ingredient products	Asulox.
Permitted mixtures	None significant.
Risks and precautions	Avoid breathing drift or dust; keep animals out of treated fields for at least 2 weeks.

Atrazine

Function	General residual herbicide [heterocyclic, triazine].
Uses	Against annual dicotyledons and grasses in maize and raspberries, in forestry, and on non-crop areas.
Single ingredient products	Ashlade, Atlas Atrazine, Borocil, Gesaprim.
Permitted mixtures	With amitrole, bromacil, cyanazine, 2,4-D, dicamba, diuron, imazapyr, MCPA, mecoprop, picloram, sodium chlorate, terbuthylazine.
Products based on mixtures	Bullseye, Boracil extra, Holtox, Arsenal, **Boots Long-Lasting Weed Killer**, Deeweed, **Murphy Path Weed Killer**, **Maskell Total Weedkiller**, **Rentokil Path & Patio Weedkiller**.
Risks and precautions	Irritates skin and eyes.
Status	Under review.

Azamethiphos

Function	Residual insecticide [organophosphorus].
Uses	For fly control in animal houses.

Single ingredient products	Alfacron.
Permitted mixtures	None significant.
Risks and precautions	Irritating to eyes and skin; can cause skin sensitisation; dangerous to fish.

Bacillus thuringiensis

Function	Insecticide [bacterial].
Uses	Against caterpillars of cabbage white butterfly and others on brassicas, raspberries, tomatoes, beans, cucumbers, and ornamentals.
Single ingredient products	Atlas Thuricide, Bactospeine, **Dipel**.
Permitted mixtures	With pyrethrins.
Products based on mixtures	**Bactospeine Garden**.

Benazolin

Function	Translocated herbicide [growth regulator].
Uses	In mixtures for post-emergence control of weeds under cereals, brassicas.
Single ingredient products	Available only in mixtures in the UK.
Permitted mixtures	With bromoxynil; clopyralid; 2,4-D; 2,4-DB; ioxynil; MCPA; mecoprop.
Products based on mixtures	Asset, Benazalox, Jaguar, Legumex Extra, Setter, **Boots Total Lawn Treatment, Fisons New Evergreen**.
Risks and precautions	Irritating to skin and eyes; harmful to fish.
Status	Review planned.

Bendiocarb

Function	Systemic and contact insecticide [carbamate].

Uses	Against pygmy beetles, springtails, wireworms in beets; bean seed flies in beans, courgettes, marrows, and squashes; household ants.
Single ingredient products	Ficam, Ficam Wasp and Hornet Killer, Garvox, Seedox, **Camco Ant Powder**, **Delta Insect Powder**, **Pyrenone Bendiocarb Wasp and Hornet Spray**.
Permitted mixtures	With pyrethrins.
Products based on mixtures	**Pyrenone Bendiocarb Roach and Ant Spray**, **Raid Ant Bait**.
Risks and precautions	Harmful if swallowed or contacts skin (anticholinesterase).

Benomyl

Function	Systemic fungicide [carbamate].
Uses	Against a wide range of fungi on cereals, rape, peas and beans, celery, lettuce, and other vegetables as well as fruit bushes and flowers.
Single ingredient products	**Benlate**.
Permitted mixtures	With iodofenphos, metalaxyl.
Products based on mixtures	Polycote Pedigree.
Status	Under review.

Bentazone

Function	Post-emergence contact herbicide.
Uses	Against annual dicotyledons in beans, linseed, potatoes, and onions.
Single ingredient products	Basagran.
Permitted mixtures	With cyanazine, 2,4-D, 2,4-DB, dichlorprop, isoproturon, MCPA, MCPB, mecoprop.
Products based on mixtures	Topshot, Acumen, Pulsar, Tripart Merso.

| Risks and precautions | Irritating to eyes. |

Bifenox

Function	Herbicide [chlorinated phenoxy].
Uses	Against grasses in cereals, usually post-emergence.
Single ingredient products	Used only in mixtures.
Permitted mixtures	With chlorotoluron, isoproturon, mecoprop.
Products based on mixtures	Dicurane Duo, Invicta, Foxstar.
Risks and precautions	Irritating to eyes and skin; harmful to fish.

Bioallethrin

Function	Insecticide [pyrethoid].
Uses	Contact, non-systemic insecticide for household pests; usually combined with synergists.
Single ingredient products	**Boots Electric Mosquito Killer**, **Buzz-Off**, **First Class Mosquito Killer Travel Pack**.
Permitted mixtures	With permethrin, tetramethrin.
Products based on mixtures	Insektigun, **Cooper Ant and Crawling Insect Powder**, **Cooper Fly and Wasp Killer**, **Floracid**, **Kill-a-Bug**.
Risks and precautions	Protect food and pets from contamination when spraying; dangerous to fish.

Bioresmethrin

| Function | Broad-contact insecticide. |
| Uses | Against cockroaches, flies, mosquitoes. |

Single ingredient products	Cooper's Garden Pest Killer.
Permitted mixtures	With bioallethrin.
Products based on mixtures	Cooper's Fly Spray, Cooper's Pybuthrin 33 BB.
Risks and precautions	Dangerous to bees and fish.

Bromacil

Function	Soil-acting herbicide [heterocyclic].
Uses	Against annual weeds in cane fruits and for total vegetation control in non-crop areas.
Single ingredient products	Borocil, Hyvar.
Permitted mixtures	With amitrole, atrazine, diuron, pentachlorophenol, picloram.
Products based on mixtures	Destral, Boracil Extra, Fenocil, Hydon.
Risks and precautions	Irritating to skin, eyes, lungs.
Status	Review planned.

Bromophos

Function	Non-specific stomach and contact insecticide [organophosphorus].
Uses	Against flies, wasps, and some caterpillars. Often mixed with fungicides.
Single ingredient products	Bromophos dust, Hybrom, Nexion, **Agrichem Caterpillar Spray**, **Agrichem Root Fly Spray**, **Nexion Oil**, **PBI Bromophos**, **Springspray**.
Permitted mixtures	With captan, thiabendazole.
Products based on mixtures	Bromotex.
Risks and precautions	Irritating to eyes, skin, and lungs (organophosphorus); dangerous to fish.

Bromoxynil

Function	Contact herbicide [benzonitrile].
Uses	Against dicotyledons in cereals.
Single ingredient products	Used in mixtures only.
Permitted mixtures	With benazolin, chlorsulfuron, clopyralid, dichlorprop, ethofumesate, fluroxypyr, ioxynil, linuron, MCPA, trifluralin.
Products based on mixtures	Advance, Alpha Briotril, Asset, Atom Certrol E, Atlas Minerva, Jaguar, Glean, Vindex, Crusader S, Bromoxolon, Banvel, Sickle Masterspray.
Risks and precautions	Harmful if swallowed or contacts skin; harmful to fish.
Status	Under review.

Bupirimate

Function	Systemic fungicide [heterocyclic].
Uses	Against powdery mildews in courgettes, cucumbers, marrows, tomatoes, cane fruits, and flowers.
Single ingredient products	Nimrod.
Permitted mixtures	With chlorothalonil, pirimicarb, triforine.
Products based on mixtures	FF 4083, Nimrod T, **Roseclear**, **Nimrod T.**
Risks and precautions	Irritating to eyes and skin; harvest intervals: 1–7 days on soft fruit and 14 days on hops.

Captan

Function	Fungicide.
Uses	Against apple scab, fungal fruit rot, black spot in roses; botrytis in flowers.
Single ingredient products	Captan.

Permitted mixtures	With bromophos, fosetyl-aluminium, thiabendazole; 4-indole-3-ylbutyric acid, lindane, malathion, metalaxyl, 1-naphthylacetic acid, nuarimol, penconazole.
Products based on mixtures	Bromotex, Aliette SD, Hy-Cote, Grammalex, Apron 70SD, Kapitol, HY-T, **Rooting Powder, Combined Seed Dressing, Keri-Root, Murphy Hormone Rooting Powder**.
Risks and precautions	Irritating to eyes, skin, and lungs.
Status	Under review.

Carbaryl

Function	Contact insecticide [carbamate].
Uses	Against a range of insects on apples, pears, soft fruits, brassicas, lettuce, peas, ornamentals; earthworms in amenity grass; mites and spiders in poultry houses.
Single ingredient products	Carbaryl 45, Carbaryl Dusting Powder, Microcarb, Scattercarb, Sevin 85 Sprayable, Sevin XLR, Tornado. **Dethlac Ant and Insect Powder, Fisons Ant Trap, Murphy Wasp Destroyer, Rentokil Ant & Insect Powder, Rentokil Wasp Nest Killer**.
Permitted mixtures	With pyrethrins, quintozene, rotenone, sodium tetraborate.
Products based on mixtures	**Secto Wasp Killer Powder, Boots Garden Insect Powder, Boots Ant Killer**.
Risks and precautions	Harmful if swallowed (anticholinesterase); dangerous to bees.
Status	Review planned.

Carbendazim

Function	Systemic fungicide [carbamate].
Uses	Against a wide range of fungus diseases on cereals, beans, tree and soft fruit, flowers. Also used in wood preservatives.
Single ingredient products	Ashlade Carbendazim Flowable, Bavistin, Carbate, Fisons Turfclear, Supercarb, Fazala

K75, **Bio One-Shot for Diseases**, **Boots Garden Fungicide**, **Murphy Zap Cap Systemic Fungicide**, **PBI Supercarb**.

Permitted mixtures	With chlorothalonil, flusilazole, flutriafol, mancozeb, maneb, metalaxyl, permethrin, prochloraz, propiconazole, sulphur, tecnazene, thiram, tributylin oxide, vinclozolin, zinc naphthenate.
Products based on mixtures	Mastiff, Bravocarb, Tripart Victor, Corbel Duo, Punch C, Early Impact, Kombat, Multi W, Ashlade M, Delsene M Flowable, Headland Dual, Septal, Squadron, Bolda FL, Cosmic FL, Brolly, Konker, **Bio Multivag**, **Dulux Weathershield Exterior Preservative Basecoat**, **Bio Woody**, **Trispot Exterior Mouldicide**, **Mystic General Garden Insecticide**.
Risks and precautions	Keep away from food and pets. Harmful to fish.
Status	Under review.

Carbofuran

Function	Systemic insecticide [carbamate].
Uses	Against a wide range of insects on beets, brassicas, swedes, potatoes, rape, carrots, maize, strawberries, and ornamentals.
Single ingredient products	Carbosip, Nex, Power Cannon, Rampart, Yaltox.
Permitted mixtures	None significant.
Risks and precautions	Harmful if swallowed or contacts skin (anticholinesterase); dangerous to livestock, fish, game, wild birds, and animals; keep livestock out of treated areas for at least 2 weeks.

Chlordane

Function	Persistent earthworm killer [organochlorine].
Uses	For amenity grass.
Single ingredient products	APC Wormkiller, Mascot Wormer, Parker's Wormex, Sydane.
Permitted mixtures	None significant.

Risks and precautions	Harmful if swallowed or contacts skin; keep all livestock out of treated areas for at least 14 days.

Chlorfenvinphos

Function	Contact and ingested soil insecticide [organophosphorus].
Uses	Against wheat bulb and yellow cereal fly in wheat; cabbage root fly in brassicas including radishes; carrot fly; Colorado beetle in potatoes.
Single ingredient products	Birlane, Sapecron.
Permitted mixtures	None significant.
Risks and precautions	Toxic if swallowed or contacts skin (organophosphorus); irritating to eyes and skin; dangerous to fish; harmful to game, wild birds, and animals.
Status	Review planned.

Chloridazon

Function	Residual herbicide [heterocyclic].
Uses	Against annual weeds in beets.
Single ingredient products	Ashlade Chloridazon, Atlas Silver, Hyzon, New Murbetex, Pyramin DF, Starter Flowable, Trojan.
Permitted mixtures	With chlorbufam, chlorpropham, di-allate, ethofumesate, fenuron, lenacil, propachlor, propham.
Products based on mixtures	Alicep, Magnum, Spectron, Barrier, Advizor, Varmit, Ashlade CP, Atlas Electrum.
Risks and precautions	Harmful if swallowed; may cause skin sensitisation; irritating to eyes, skin, and lungs.
Status	Review planned.

Chlormequat

Function	Plant growth regulator.
Uses	For lodging control in cereals.

Single ingredient products	ABM Chlormequat, Cleanacres, Farmacel, Hyquat, Titan, Tripart Brevis.
Permitted mixtures	With 2-chloroethylphosphonic acid, choline chloride, di-l-p-menthene.
Products based on mixtures	Pacer, Upgrade, Arotex Extra, Mandops Halloween, Mandops Hele-Stone, Portman Superquat.
Risks and precautions	Harmful if swallowed or contacts skin.

Chlorophacinone

Function	Rodenticide [anticoagulant].
Uses	Against voles, rats, and mice in buildings.
Single ingredient products	Drat, Drat Rat and Mouse Bait, Karate, Ridento Ready-To-Use-Bait, Sakarat Special.
Permitted mixtures	None significant.
Risks and precautions	Harmful if swallowed or contacts skin; keep away from animals and children.
Status	Review planned.

Chlorothalonil

Function	Fungicide.
Uses	Against a wide range of fungi on wheat, potatoes, peas and beans, brassicas, celery, onions and leeks, soft fruit, tomatoes, amenity grass, and glasshouse crops like flowers, cucumbers, and peppers. Also used in paints and adhesives.
Single ingredient products	Ashlade Bravo, Bombardier, Bravo 500, Comet, Daconil, Jupital, Repulse, Tripart Faber.
Permitted mixtures	With bupirimate, carbendazim, fenpropimorph, flutriafol, maneb, metalaxyl, propiconazole.
Products based on mixtures	Bravocarb, Corbel CT, Impact Excel, Mistral, Folio 575 FW, Sambarin TP.
Risks and precautions	Irritating to eyes, lungs, and skin; can cause allergy, with redness of eyes, bronchial irritation, and skin rash; dangerous to fish; harvest interval: 12 hours for tomatoes, peppers, cucumbers;

| | 3 days for soft fruits; 7 days for celery and brassicas; 10 days for beans, peas, and onions. |
| Status | Under review. |

Chlorotoluron

Function	Contact and residual herbicide [amides and ureas].
Uses	Against blackgrass, wild oats, annual grasses, and dicotyledons in wheat, barley, durum wheat, and triticale.
Single ingredient products	Dicurane, Hyvena S, Ludorum, Phyto-Toluron.
Permitted mixtures	With bifenox, tri-allate.
Products based on mixtures	Dicurane Duo.
Risks and precautions	Irritating to skin and eyes.

Chloroxuron

Function	Residual herbicide [amides and ureas].
Uses	Against annual dicotyledons and meadow grass in strawberries, herbaceous ornamentals; moss on hard surfaces.
Single ingredient products	Tenoran.
Permitted mixtures	With dichlorophen, ferrous sulphate.
Products based on mixtures	Ashlade D-Moss, **Mosskiller**, **Murphy Tumblemoss**.
Status	Review planned.

Chlorpropham

| Function | Residual herbicide and sprout suppressant [carbamate]. |
| Uses | Against annual grasses and dicotyledons, especially chickweed and polygonums in onions, leeks, carrots, lettuce, celery, parsley, |

and soft fruits; and for sprout suppression in potatoes.

Single ingredient products	Atlas CIPC 40, Mirvals 500 HN, MSS CIPC, Triherbicide CIPC, Warefog 25.
Permitted mixtures	With cetrimide, chloridazon, cresylic acid, 2,4-D, diuron, fenuron, linuron, maleic hydrazide, pentanochlor, propham.
Products based on mixtures	Atlas Brown, Atlas Gold, Atlas Herbon Yellow, Atlas Pink, Atlas Red, Croptex Pewter, MSS Sugarbeet Herbicide, Pommetrol M, Power Gro-Stop, Profalon, Profen.
Risks and precautions	Harmful if swallowed, breathed, or contacts skin; irritating to eyes, lungs, and skin; potatoes must not be sold within 21 days of treatment.
Status	Review planned.

Chlorpyrifos

Function	Contact and ingested insecticide [organophosphorus].
Uses	Against a wide range of insects on cereals, amenity grass, brassicas, potatoes and other vegetables, strawberries, rhubarb, tree fruit, and in forestry.
Single ingredient products	Crossfire, Dursban, Spannit, Talon, Killgerm Terminate. **Murphy Super Root Guard**.
Permitted mixtures	With dimethoate, disulfoton.
Products based on mixtures	Atlas Sheriff, Twinspan. **Chlorophos**.
Risks and precautions	Harmful if swallowed or contacts skin (organophosphorus); irritating to eyes and skin; dangerous to bees and fish; harvest intervals: 7 days for soft fruit; 14–21 days for most other vegetable and fruit crops; 6 weeks for brassicas.

Chlorpyrifos-methyl

Function	Insecticide and acaricide [organophosphorus].
Uses	Against grain-storage pests.
Single ingredient products	Cooper Graincote, Cooper Smite Liquid, Redlan 50.

Permitted mixtures	With permethrin, pyrethrins.
Products based on mixtures	Cooper Multispray.
Risks and precautions	Harmful if swallowed or contacts skin (organophosphorus); irritating to eyes and skin; dangerous to fish.

Chlorthal-dimethyl

Function	Residual herbicide [amides and ureas].
Uses	Against annual dicotyledons in brassicas, beans, onions, and leeks, soft fruits, ornamentals, and amenity grass.
Single ingredient products	Dacthal.
Permitted mixtures	With propachlor.
Products based on mixtures	Decimate.
Risks and precautions	Irritating to eyes and skin.

Clopyralid

Function	Systemic herbicide [heterocyclic].
Uses	Against dicotyledons like mayweed, corn marigold, creeping thistle in beets, swedes, rape, cereals, maize, brassicas, grassland, hardy nursery ornamentals.
Single ingredient products	Dow Shield.
Permitted mixtures	With benazolin, bromoxynil, cyanazine, dichlorprop, fluroxypyr, ioxynil, MCPA, mecoprop, phenmedipham, propyzamide, triclopyr.
Products based on mixtures	Benazalox, Broxolon, Campaign, Coupler SC, Crusader, Escort, Grazon, Harrier, Hotspur, Lontrel Plus, Seloxone, Matrikerb, Vindex.
Risks and precautions	Irritating to eyes; keep off skin and hands.

Copper hydroxide

Function	Fungicide and bactericide.
Uses	Against diseases like leaf spot in cereals; halo blight in beans; downy mildew in brassicas, grape vines, and onions; potato blight; apple canker; leaf curl in apricots and peaches; leaf spots in soft fruits.
Single ingredient products	Chiltern Kocide, Comac Parasol, Wetcol.
Permitted mixtures	None significant.
Risks and precautions	Harmful if swallowed; irritating to eyes, lungs, and skin; keep all livestock out of treated area for at least 3 weeks.
Status	Review planned.

Copper oxychloride

Function	Fungicide and bactericide.
Uses	Against blight in potatoes and outdoor tomatoes; celery leaf spot; canker in cherries and plums; downy mildew in grape vines; leaf curl in apricots and nectarines; collar rot in apples; and collar rot in calabrese.
Single ingredient products	Cuprokylt, Cuprosana, FS Dricol 50, **Murphy Traditional Copper Fungicide**.
Permitted mixtures	With carbendazim, maneb, metalaxyl, permethrin, sulphur.
Products based on mixtures	Ashlade SMC, Tripart Senator Flowable, Ridomil, **Bio Multiveg**.
Risks and precautions	Irritating to eyes, lungs, and skin; avoid breathing fumes or dust; keep all livestock out of treated area for at least 3 weeks.
Status	Review planned.

Copper sulphate

Function	Fungicide.
Uses	Usually in mixtures (Bordeaux mixture) against blight in potatoes and tomatoes; canker in

apples, pears, and cherries; leaf curl in peaches, nectarines, and apricots; leaf spots in soft fruits; powdery mildews in grape vines and hops.

Permitted mixtures	With aluminium sulphate, ammonium carbonate, ammonium hydroxide, cufraneb, ferrous sulphate, sodium hydrogen carbonate, sodium tetraborate, sulphur.
Products based on mixtures	Comac Macuprax, Corry's Moss Remover, Top-Cop, **Bio Multiveg**, **Cheshunt Compound**, **Bordeaux Mixture**, **Noble Garden Pack**, **Spraydex Fungicide**.
Risks and precautions	Harmful if swallowed; irritating to eyes, lungs, and skin; dangerous to fish.
Status	Review planned.

Coumatetralyl

Function	Anticoagulant rodenticide.
Uses	Against rats and mice in buildings.
Single ingredient products	Racumin bait, Racumin Tracking Powder, Townex Rat Bait, **PBI Racumin Mouse/Rat Bait**.
Permitted mixtures	None significant.
Risks and precautions	Harmful if swallowed; keep away from children and animals.

Cresylic acid

Function	Contact fungicide, soil sterilant, insecticide.
Uses	Against honey fungus, canker, crown gall in trees and shrubs; overwintering pests in fruit trees; ants, slugs, and woodlice in glasshouses.
Single ingredient products	Armillatox, Bray's Emulsion, **Clean Up**, **Synchemicals Medo**.
Permitted mixtures	None significant.
Risks and precautions	Irritating to eyes, lungs, and skin; dangerous to fish.

Cupric ammonium carbonate

Function	Fungicide.
Uses	Against leaf mould in tomatoes; powdery mildew in chrysanthemums and cucumbers; leaf spot in celery and soft fruits; damping off in ornamental seedlings.
Single ingredient products	Croptex Fungex.
Permitted mixtures	None significant.
Risks and precautions	Harmful if swallowed.

Cyanazine

Function	Contact and residual herbicide [heterocyclic, triazine].
Uses	Against annual dicotyledons and grasses in cereals, rape, peas and beans, potatoes, maize, and brassicas.
Single ingredient products	Fortrol.
Permitted mixtures	With atrazine, bentazene, clopyralid, 2,4-DB, fluroxypyr, isoproturon, mecoprop.
Products based on mixtures	Cleval, Holtox, Quiver, Spitfire, Topshot.
Risks and precautions	Harmful if swallowed or contacts skin; irritating to eyes.

Cyfluthrin

Function	Contact and residual insecticide [pyrethroid].
Uses	Against aphids in cereals; cabbage stem flea beetles in rape; caterpillars, aphids, flea beetles in brassicas; weevils in peas and beans; apple moths, capsids, and midges in soft fruits.
Single ingredient products	Baythroid
Permitted mixtures	None significant.

Risks and precautions	Harmful if swallowed; irritating to eyes and skin; dangerous to fish and bees.

Cymoxanil

Function	Systemic fungicide [amides and ureas].
Uses	Against potato blight.
Single ingredient products	Available in mixtures only.
Permitted mixtures	With mancozeb, oxadixyl.
Risks and precautions	Irritating to eyes, lungs, and skin.

Cypermethrin

Function	Contact and stomach-acting insecticide [pyrethroid].
Uses	Against a wide range of insect pests in cereals, vegetables, fruits, ornamentals. Also in wood treatments.
Single ingredient products	Kordon Insecticidal Lacquer, Vapona Antpen, Vapona Insectipen, Ambush C, Ashlade Cypermethrin, Cyperkill, Cymperator, Demon 40, Devatern, Folcord, Killgerm Precor ULV III, Quadrangle Cyper, Toppel, Vassgro Cypermethrin Insecticide, **Cementone Woodworm Killer**, **Green Range Woodworm Killer**.
Permitted mixtures	With bioborate, methoprene, permethrin, tetramethrin.
Products based on mixtures	KO Longlast, Vapona Ant and Crawling Insect Spray, **Green Range Dual Purpose AQ**, **Shelltox Cockroach and Crawling Insect Killer**.
Risks and precautions	Harmful if swallowed or contacts skin; irritates eyes and skin; can cause skin sensitisation; extremely dangerous to fish; dangerous to bees.
Status	Under review.

2,4-D

Function	Translocated herbicide [chlorinated phenoxy].
Uses	Against annual grasses and dicotyledons in cereals, grassland, turf, conifer plantations in forestry.
Single ingredient products	Agricorn, Atlas 2,4-D, Campbell's Destox, Campbell's Dioweed, Dicotox Extra, Dormone, Mascot Showerproof Selective, Planotox, Silvapron, Syford, Weedone LV4, **Double H Lawn Feed and Weed, Goodlife Lawn Weed and Feed, Notcutts Lawn Weed and Feed**.
Permitted mixtures	With amitrole, atrazine, benazolin, dicamba, dichlorprop, diuron, ferrous sulphate, ioxynil, MCPA, mecoprop, picloram, simazine, sodium chlorate, 2,4,5-T, triclopyr.
Products based on mixtures	Atlavar, Broadshot, Deltanox, Destral, Hytrol, Longlife Plus, Mascot Selective CDA, New Formulation Weed and Brush Killer, Renovator, Setter 33, Supertox 30, Super Verdone, Tordon 101 Weedkill, Weed and Feed with Didin, Zennapron, **Asda Garden Lawn Weed and Feed + Mosskiller, Boots Kill-A-Weed, Boots Lawn Weedkiller, Boots Total Lawn Treatment, Hytrol, Bio Lawn Weed Killer, Doff Lawn Weedkiller, Fisons Lawn Spot Weedkiller, Gen Lawn Weed and Feed, Greensward, Green Up Weedfree Lawn Weedkiller, Murphy Lawn Weedkiller and Lawn Tonic, Mystic Lawn Killer, PBI Toplawn, Spot Weeder, Tesco Lawn Feed'N'Weed, Touchweeder, Triple Action Grass Hopper, Wilco Lawn Weedkiller**.
Risks and precautions	Harmful if swallowed or contacts skin; irritating to eyes.
Status	Under review.

Dalapon

Function	Translocated herbicide [halogenated hydrocarbon].
Uses	Against grasses in beets, rape, potatoes, carrots, tree and soft fruits; in non-crop areas, ditches, banks and watercourses, road verges, parks.

Single ingredient products	Atlas Dalapon, Synchemicals Dalapon, **Dalapon Herbicide**, **Synchemicals Couch and Grass Killer**.
Permitted mixtures	With dichlobenil, di-l-p-menthene, divron.
Products based on mixtures	Fydulan, Volunteered, Destral.
Risks and precautions	Irritating to eyes, lungs, and skin.

Daminozide

Function	Plant growth regulator.
Uses	To shorten flowering ornamentals like chrysanthemums, poinsettias, bedding plants, and pot plants.
Single ingredient products	Alar, Dazide.
Permitted mixtures	None significant.
Risks and precautions	Irritating to eyes.
Status	UK approved but manufacturers have withdrawn the product for food-related uses.

Dazomet

Function	Soil fumigant.
Uses	Against weed seeds and soil-living diseases; organisms like nematodes and insects in field crops, vegetables, glasshouses, loam, and compost.
Single ingredient products	Basamid, Bosamet, Dazomet.
Permitted mixtures	None significant.
Risks and precautions	Harmful if swallowed or contacts skin; irritating to eyes, lungs, and skin.
Status	Review planned.

2,4-DB

Function	Systemic herbicide [chlorinated phenoxy].

Uses	Against dicotyledons in lucerne, and in cereals planted in a lucerne crop.
Single ingredient products	Atlas Pastral, Campbell's DB Straight, Embutox 30, Marks 2,4-DB.
Permitted mixtures	With benazolin, bentazone, cyanazine, linuron, MCPA, mecoprop.
Products based on mixtures	Alistell, Clovacorn Extra, Campbell's Bellclo, Embutox Plus, Farmon 2,4-DB Plus.
Risks and precautions	Harmful if swallowed or contacts skin.
Status	Review planned.

Deltamethrin

Function	Contact and residual insecticide [pyrethroid].
Uses	Against a wide range of insects like aphids, cabbage stem flea beetle, codling moth in field crops, soft fruits, tree fuits, ornamentals, nursery stock.
Single ingredient products	Crackdown, Decis, Thripstick.
Permitted mixtures	With heptenophos.
Products based on mixtures	Desiquick.
Risks and precautions	Harmful if swallowed or contacts skin; irritating to eyes and skin; extremely dangerous to fish and dangerous to bees.
Status	Under review.

Demeton-S-methyl

Function	Systemic and contact insecticide [organophosphorus].
Uses	Against a wide range of insect pests like aphids, thrips, leafhoppers, red spider mites in a wide range of crops including field vegetables, cereals, soft fruits, tree fruits, flowers and ornamentals, and glasshouse vegetables.
Single ingredient products	Ashlade Persyst, Campbell's DSM, Chafer Azotox, Metasystox 55, Mifatox, Power DSM

	580, Tripart Systemic Insecticide, Vassgro DSM.
Permitted mixtures	None significant.
Risks and precautions	Toxic if swallowed, breathed, or contacts skin (organophosphorus); harmful to bees, game, birds, and fish.
Status	Under review.

Demeton-S-methyl sulphone

Function	Insecticide [organophosphorus].
Uses	Against a wide range of insects like aphids, caterpillars and moths, red spider mites, and Colorado beetles in crops like tree fruits, soft fruits, brassicas, potatoes, nursery stock.
Single ingredient products	Used only in mixtures.
Permitted mixtures	With azinphos-methyl.
Products based on mixtures	Gusathion.
Risks and precautions	Toxic if swallowed, breathed, or contacts skin (organophosphorus); dangerous to bees, birds, game, and fish; keep all livestock out of treated areas for at least 2 weeks; harvest interval: 3 weeks.
Status	Under review.

2,4-DES

Function	Herbicide [chlorinated phenoxy].
Uses	Against dicotyledons and annual grasses in maize, vegetables like asparagus and rhubarb, soft fruits, nursery stock, and flower crops like roses, gladioli, narcissi, perennial flowers (in England and Wales only).
Single ingredient products	Available only in mixtures.
Permitted mixtures	With simazine.
Products based on mixtures	Atlas Herbon Blue.

Risks and precautions	Irritating to skin, lungs, and eyes; dangerous to fish.
Status	Approval withdrawn autumn 1990.

Desmetryn

Function	Contact herbicide [heterocyclic, triazine].
Uses	Against annual dicotyledons in brassicas.
Single ingredient products	Semeron 25 WP.
Permitted mixtures	None significant.
Status	Review planned.

Diazinon

Function	Insecticide [organophosphorus].
Uses	Against insects like cabbage root fly, carrot fly, leaf miner and aphid in lettuce; onion fly, leaf miner, scale insect, white fly, and red spider mite in ornamentals, flowers, tomatoes, and cucumbers.
Single ingredient products	Basudin, Darlingtons Diazinon Granules or Liquid, **Acclaim Plus**, **Cromopest Crawling Insect Killer**, **Dethlac Insecticidal Lacquer**, **Doff Antlak**, **Murphy Root Guard**, **Secto Ant & Crawling Insect Spray**, **Secto Kill-a-Line**.
Permitted mixtures	With pyrethrins.
Products based on mixtures	**Ant Gun**.
Risks and precautions	Harmful if swallowed, breathed, or contacts skin (organophosphorus); harmful to game, wild birds, and animals; dangerous to bees.
Status	Under review.

Dicamba

Function	Systemic herbicide [chlorinated phenoxy].
Uses	Against bracken in forestry, amenity grass, and non-crop areas; against perennial dicotyledons on grassland.

Single ingredient products	Tracker.
Permitted mixtures	With benazolin, 2,4-D, dichlorophen, dichlorprop, ferrous sulphate, maleic hydrazide, MCPA, mecoprop, paclobutrazol, triclopyr.
Products based on mixtures	Banlene Plus, Campbell's Field Marshal, Chafer Mephetol Extra, Di-Farmon, Dock-Ban, Docklene, BH Dockmaster, Endox, Fettel, Headland Relay, Herrisol, Holdfast, Hyban, Hygrass, Hyprone, Hysward, Mascot Super Selective, Mazide Selective, Paddox, Pasturol, Quadban, Selectrol, Tribute, Tritox, **Bio Lawn Weed Killer, Bio Weed Pencil, Boots Nettle and Bramble Weedkiller, Boots Total Lawn Treatment, Fisons Lawncare Liquid, Fisons Lawn Spot Weedkiller, Green Up Lawn Feed And Weed, Green Up Weedfree Lawn Weedkiller, Greensward, Lawnsman Weed and Feed, Lawnweed Gun, Mystic Lawn Weedkiller, Oak Lawnmaster Lawn Feed With Weedkiller, PBI Toplawn, Tesco Lawn Feed'N'Weed, Triple Action Grass Hopper, Velvetone Lawn Food + Weedkiller.**
Risks and precautions	Keep livestock out of treated area for at least two weeks after treatment.
Status	Review planned.

Dichlobenil

Function	Residual herbicide [benzonitrile].
Uses	Against annual and perennial weeds in soft fruits and established trees.
Single ingredient products	BH Prefix D, Casoron G, Prefix D.
Permitted mixtures	With dalapon.
Risks and precautions	Harmful to poultry.
Status	Review planned.

Dichlofluanid

Function	Fungicide [chlorinated phenoxy].
Uses	Against botrytis, cane spot, and mildew in cane fruit; downy mildew in cauliflower; leaf mould in tomatoes; black spot in roses. Also used in wood preservatives.
Single ingredient products	Elvaron, **Blue Peter Intertox, Cuprinol Hardwood Basecoat, Cedarwood Protector**.
Permitted mixtures	With acypetacs-zinc, biborate, furmecyclox, lindane, permethrin, tributylin oxide, zinc naphthenate in wood treatments.
Products based on mixtures	**ABL Wood Preservative D, Barretine New Universal Fluid, Blue Peter Double Action Wood Preservative, Cuprinol Decorative Wood Preserver, Cuprinol Preservative-Wood Hardener, Double Action Wood Preservative, Fungal Primer, Impra-colour, Gori 22, Pro-Am Timber Treat Cedar, Signpost Wood Preserver Red Cedar, Timbrene, Weathershield Preservative Primer**.
Status	Review planned.

Dichlorophen

Function	Moss-killer and algicide [chlorinated phenoxy].
Uses	Against mosses and algae in paths, turf, and hard surfaces; mosses and liverwort on pot plants, slime moulds in water pipes.
Single ingredient products	Algofen, Fungo, Mascot Moss Killer, Panacide M, Super Mosstox, **Bio Moss Killer, Boots Spring and Autumn Lawn Treatment, Mossgun, Mosstox Green, Mosstox Plus**.
Permitted mixtures	With benazolin, chloroxuron, 2,4-D, dicamba, dichlorprop, ferrous sulphate, mecoprop.
Products based on mixtures	**Boots Total Lawn Treatment, Mosskiller, J. Arthur Bower's Mosskiller**.
Status	Review planned.

Dichlorprop

Function	Systemic herbicide [chlorinated phenoxy].
Uses	Against annual dicotyledons in cereals.
Single ingredient products	Campbell's Redipon, MSS 2,4-DP, Cleanacres 2,4-D, Marks Polytox-K/M, Tripart 2,4-DP 60.
Permitted mixtures	With bentazene, bromoxynil, clopyralid, 2,4-D, dicamba, dichlorophen, ferrous sulphate, MCPA, mecoprop.
Products based on mixtures	SHL Turf Feed and Weed + Weedkiller, Campbell's Redipon Extra, Farmon 2,4-DP + MCPA, Seritox 50, Seritox Turf, **Boots Kill-a-Weed, Boots Lawn Weedkiller, Boots Nettle and Bramble Weedkiller, Doff Lawn Spot Weeder, Doff Lawn Weedkiller, Garden Centre Lawn Feed, Weed and Moss Killer, Lawnsman Liquid Weed and Feed, Murphy Lawn Weedkiller and Lawn Tonic, Wilko Lawn Weedkiller.**
Risks and precautions	Harmful if swallowed or contacts skin.
Status	Review planned.

Dichlorvos

Function	Contact and fumigant insecticide [organophosphorus].
Uses	Against flies, mosquitoes, beetles, and ectoparasites in poultry houses; general insects in glasshouses; western flower thrips in non-edible glasshouse crops.
Single ingredient products	Darmycel Dichlorvos, Nuvan 500 EC, **Fix Up Slow Release Fly Killer.**
Permitted mixtures	With fenitrothion, lindane, permethrin, propoxur, pyrethrins, tetramethrin.
Products based on mixtures	Blattanex Residual Spray, **Keen Superkill Ant & Roach Exterminator, Nuvan Fly Killer, Secto Aphid Killer, Secto Household Flea Killer.**
Risks and precautions	Toxic if swallowed, breathed, or contacts skin (organophosphorus); keep unprotected people

out of treated area for at least 12 hours;
dangerous to bees and fish.

Status	Under review.

Diclofop-methyl

Function	Systemic herbicide [chlorinated phenoxy].
Uses	Against annual grasses in cereals, field beans, brassicas, lettuce, rape, onions, parsnips, and potatoes.
Single ingredient products	Hoegrass.
Permitted mixtures	None significant.
Risks and precautions	Irritating to skin and eyes; keep livestock out of treated area for at least 7 days.

Dicloran

Function	Fungicide [amides and ureas].
Uses	Against botrytis and rhizoctonia in tomatoes, lettuce, cucumbers, glasshouse flowers.
Single ingredient products	Fumite Dicloran Smoke.
Permitted mixtures	With thiram.
Products based on mixtures	Tubair Botryticide.
Risks and precautions	Irritating to eyes and lungs; harvest intervals: cucumbers and tomatoes 2 days, lettuce 14 days.
Status	Review planned.

Difenzoquat

Function	Post-emergence herbicide [heterocyclic].
Uses	To control wild oats in cereals.
Single ingredient products	Avenge 2.
Permitted mixtures	None significant.
Risks and precautions	Harmful if swallowed or contacts skin; irritating to skin; harmful to eyes.

Diflubenzuron

Function	Contact and stomach insecticide [amides and ureas].
Uses	Against caterpillars on brassicas, fruit trees, soft fruits, tomatoes, peppers, ornamentals.
Single ingredient products	Antec Larvakill, Dimilin.
Permitted mixtures	None significant.

Diflufenican

Function	Shoot-absorbed herbicide.
Uses	Against broadleaved weeds and grasses in cereals.
Single ingredient products	Available only in mixtures.
Permitted mixtures	With isoproturon.
Products based on mixtures	Cougar, Javelin, Panther.

Dimethoate

Function	Contact and systemic insecticide and acaricide [organophosphorus].
Uses	Against a wide range of insects on cereals, brans, beets, brassicas, parsnips, celery, potatoes, tomatoes, leeks, onions, soft fruits, fruit trees, ornamentals, flowers.
Single ingredient products	Atlas Dimethoate, Campbell's Dimethoate, Tubair Systemic Insecticide, **Bio Systemic Insecticide**, **Boots Greenfly and Blackfly Killer**, **Doff Systemic Insecticide**, **Keriguards**, **Murphy Systemic Insecticide**.
Permitted mixtures	With chlorpyrifos, lindane, malathion, permethrin.
Products based on mixtures	Atlas Sheriff, Sheriff, **Bio Long Last**, **Fisons Greenfly and Blackfly Killer**, **Flora Spray Systemic Insect Killer**, **Secto Garden Powder**, **Secto Rose and Flower Spray**.

Risks and precautions	Harmful if swallowed, breathed, or contacts skin (organophosphorus); harmful to game, wild birds, and animals; dangerous to bees; keep all livestock out of treated areas for at least 7 days; harvest intervals: lettuce 4 weeks, watercress 40 days, other crops 7 days.
Status	Under review.

Dinocap

Function	Fungicide [chlorinated phenoxy].
Uses	Against powdery mildews in fruit, hops, cucumbers, flowers.
Single ingredient products	Campbell's Dinocap 50% Miscible, Karthane Liquid, **Bio Multirose**.
Permitted mixtures	With permethrin, sulphur, triforine.
Risks and precautions	Harmful if swallowed, breathed, or contacts skin; irritating to eyes and lungs; dangerous to fish.
Status	Under review.

Diquat

Function	Contact herbicide [heterocyclic].
Uses	Against pre-harvest desiccation of potatoes, rape, peas, and linseed; annual dicotyledons in navy beans, mustard, sunflowers, ornamentals, borage, grape vines.
Single ingredient products	Midstream, Power Diquat, Reglone.
Permitted mixtures	With amitrole, paraquat, simazine.
Products based on mixtures	Farmon, Parable, Soltair, **Pathclear**, **Weedol**.
Risks and precautions	Harmful if swallowed; irritating to skin and eyes; do not feed treated material to animals within 4 days of spraying; keep livestock out of treated areas for at least 24 hours after treatment; do not drink treated water within 24 hours.
Status	Review planned.

Disulfoton

Function	Systemic aphicide and insecticide [organophosphorus].
Uses	Against aphids in brassicas, parsnips, beets, french beans, marrows, parsley, strawberries; carrot fly and bulb fly in narcissi.
Single ingredient products	Campbell's Disulfoton, Disyston, Solvigran.
Permitted mixtures	With chlorpyrifos, fonofos, quinalphos.
Products based on mixtures	Doubledown, Knave, Twinspan.
Risks and precautions	Toxic if swallowed, breathed, or contacts skin (organophosphorus); dangerous to game, wild birds, animals, and fish.
Status	Review planned.

Diuron

Function	Residual herbicide [amides and ureas].
Uses	For total weed control on non-crop areas; annual grasses and dicotyledons under fruit trees and soft fruit bushes; around trees and nursery stock.
Single ingredient products	Chipko Diuron 80, Diuron 80, Hoechst Diuron, Karmex, Promark Diuron, Unicrop Flowable Diuron.
Permitted mixtures	With amitrole, atrazine, bromacil, chlorpropham, 2,4-D, dalapon, MCPA, propham, simazine, tar oils.
Products based on mixtures	Atlas Pink, Borocil, Bullseye, Destral, Dexuron, Duranox, Kagolin, Krovar, Mascot Pre-M-Total CDA, Simfix Granules, Trix, Weedazol Total, **Hytrol**, **Total Weedkiller Granules**.
Risks and precautions	Harmful if swallowed or contacts skin; irritating to eyes, lungs, and skin.
Status	Review planned.

Dodemorph

Function	Systemic fungicide.

Uses	Against powdery mildew in roses, ornamental nursery stock, nursery fruit trees and bushes.
Single ingredient products	F238.
Permitted mixtures	None significant.
Risks and precautions	Harmful if contacts skin; irritating to eyes, lungs, and skin.

Endosulfan

Function	Contact and ingested insecticide and acaricide [organochlorine].
Uses	Against pollen beetles, midges, and seed weevils in rape and mustard; mites in blackcurrants, western flower thrips in non-edible glasshouse ornamentals; Colorado beetles in potatoes.
Single ingredient products	Thiodan 20.
Permitted mixtures	None significant.
Risks and precautions	Toxic if swallowed; harmful if contacts skin; harvest intervals: blackcurrants, strawberries, blackberries 6 weeks; hops, oilseed rape 3 weeks; ornamentals 24 hours.
Status	Review planned.

EPTC

Function	Residual herbicide [carbamate, thiocarbamate].
Uses	Against annual grasses and dicotyledons like chickweed and speedwell in potatoes.
Single ingredient products	Eptam 6E.
Permitted mixtures	None significant.
Risks and precautions	Harmful if swallowed; irritating to skin and eyes.

Ethirimol

Function	Systemic fungicide [heterocyclic].

Uses	Against powdery mildews in cereals.
Single ingredient products	Milgo, Milstem.
Permitted mixtures	With flutriafol, fuberidazole, imazalil, thiabendazole, triadimenol.
Products based on mixtures	Ferrax.
Risks and precautions	Irritating to eyes.

Ethofumesate

Function	Herbicide.
Uses	Against annual dicotyledons and meadow grass; blackgrass in beets, strawberries, grass-seed crops, established grassland.
Single ingredient products	Norton, Power Ethofumesate.
Permitted mixtures	With phenmedipham.
Products based on mixtures	Betanal Tandem.

Ethoprophos

Function	Nematicide and insecticide [organophosphorus].
Uses	Against potato cyst nematode.
Single ingredient products	Mocap.
Permitted mixtures	None significant.
Risks and precautions	Harmful if swallowed or breathed (organophosphorus); dangerous to livestock; keep all livestock out of treated fields for at least 2 weeks.

Etridiazole

Function	Soil-acting fungicide [heterocyclic, thiazole].
Uses	Against phytophthora in nursery-grown

ornamentals; damping off in vegetable seedlings;
fungi in strawberries and tomatoes.

Single ingredient products	Aaterra WP.
Permitted mixtures	None significant.
Risks and precautions	Irritating to eyes and skin.

Etrimfos

Function	Contact insecticide [organophosphorus].
Uses	Against grain-storage insects.
Single ingredient products	Satisfar, Satisfar Dust.
Permitted mixtures	None significant.
Risks and precautions	Toxic if swallowed, breathed, or contacts skin (organophosphorus); harmful to game, wild birds, and animals; dangerous to fish.

Fenarimol

Function	Systemic fungicide [heterocyclic].
Uses	Against mildews on soft fruits, ornamentals, roses, cucumbers; scab in apples and pears.
Single ingredient products	Rubigan, **Murphy Zap Cap Shrub and Fruit Care**.
Permitted mixtures	With permethrin.
Products based on mixtures	**Murphy Zap Cap Combined Insect Killer and Fungicide**.
Risks and precautions	Dangerous to fish.
Status	UK approved.

Fenbutatin oxide

Function	Contact and ingested acaricide.
Uses	Against red spider mite in glasshouse crops, cucumbers, tomatoes, strawberries, ornamentals.
Single ingredient products	Torque.

Permitted mixtures	None significant.
Risks and precautions	Irritating to eyes, lungs, and skin; extremely dangerous to fish.
Status	Under review.

Fenitrothion

Function	Broad-spectrum insecticide [organophosphorus].
Uses	Against a wide range of insects like leatherjackets, thrips, and frit fly in cereals, maize, and stored grains; aphids and moths in fruit trees and soft fruits.
Single ingredient products	Cooper Fenitrothion, Dicofen, Insectigas, Micromite, Rentokil Fenitrothion Dusting Powder, Sectacide 50 EC, Unicrop Fenitrothion, **Doff Garden Insect Powder**, **PBI Fenitrothion**.
Permitted mixtures	With dichlorvos, lindane, permethrin, resmethrin, tetramethrin.
Products based on mixtures	Killgerm Tetracide Insecticidal Spray, Tubair Grain Store Insecticide, **Big D Rocket Flykiller**, **Crown Crest Fly Spray**, **Keen Insect Killer**, **Keen Superkill Ant & Roach Exterminator**, **Super Raid Insecticide**, **Tox Exterminating Fly and Wasp Killer**.
Risks and precautions	Harmful if swallowed or contacts skin (organophosphorus); irritating to skin and eyes; harmful to game, wild birds and animals; dangerous to bees; harvest intervals: raspberries 7 days, other crops 2 weeks.

Fenpropimorph

Function	Contact and systemic fungicide.
Uses	Against mildews and rusts in cereals and field beans.
Single ingredient products	Corbel, Mistral, Power Fenpropimorph.
Permitted mixtures	With carbendazim, chlorothalonil, iprodione, lindane, metalaxyl, prochloraz, propiconazole, thiram, tridemorph.

Products based on mixtures	Corbel Duo, Glint 425 EC, Mistral CT, Lindex-Plus FS Seed Treatment, Sprint.
Risks and precautions	Harmful if breathed; irritating to eyes, lungs, and skin; dangerous to fish; harvest intervals: leeks 3 weeks, cereals and field beans 5 weeks.

Fentin acetate

Function	Fungicide.
Uses	Against blight in potatoes.
Single ingredient products	Available only in mixtures.
Permitted mixtures	With maneb.
Products based on mixtures	Brestan, Hytin, Trimastan.
Risks and precautions	Harmful if swallowed or contacts skin; irritating to skin and eyes; harmful to livestock; keep all livestock out of treated areas for at least 1 week; harvest interval: 1 week.
Status	Under review.

Fentin hydroxide

Function	Protectant fungicide.
Uses	Against blight in potatoes.
Single ingredient products	Ashlade Flotin, Chiltern Super-Tin 4L, Du-Ter 50, Farmatin 560, Quadrangle Super-Tin 4L.
Permitted mixtures	With maneb, metoxuron, zinc.
Products based on mixtures	Chiltern Tinman, Endspray.
Risks and precautions	Harmful if swallowed, breathed, or contacts skin; irritating to eyes, lungs, and skin; sprayed beet crops must not be fed to animals after seed harvest; harmful to livestock; all livestock to be kept out of treated areas for at least 1 week.
Status	Under review.

Fenvalerate

Function	Contact insecticide [pyrethroid].

Uses	Against aphids in cereals, brassicas, beets, apples, hops; cabbage stem moths, weevils, and pear sucker in field beans and fruit trees.
Single ingredient products	Sumicidin.
Permitted mixtures	None significant.
Risks and precautions	Harmful if swallowed or contacts skin; irritating to eyes and skin; harmful to bees; extremely dangerous to fish.

Ferrous sulphate

Function	Herbicide.
Uses	Against moss in turf.
Single ingredient products	Greenmaster Autumn, Greenmaster Mosskiller, Green-Up Mossfree, Taylor's Lawn Sand, Vitax Micro Gran 2, Vitax Turf Tonic.
Permitted mixtures	With ammonium sulphate, chloroxuron, copper sulphate, 2,4-D, dicamba, dichlorophen, MCPA, mecoprop.
Products based on mixtures	Aitkin's Lawn Sand Plus, Renovator, Eclipsall Turf Compound, **Asda Garden Lawn Weed and Feed + Mosskiller, Fisons Evergreen Extra, Gem Lawn Weed and Feed + Mosskiller, Green Up Feed and Weed Plus Mosskiller, Greensward, Mosskiller, J. Arthur Bower's Mosskiller, PBI Velas, Supergreen Feed, Weed and Mosskiller, Weed'N'Feed Extra, Wilko Lawn Feed, Weed and Mosskiller.**

Flamprop-M-isopropyl

Function	Translocated herbicide [amides and ureas].
Uses	Against wild oats in cereals.
Single ingredient products	Commando, Gunner, Power Flamprop.
Permitted mixtures	None significant.
Risks and precautions	Irritating to eyes and skin.

Fluazifop-P-butyl

Function	Herbicide [chlorinated phenoxy].
Uses	Against grasses in beets, rape, carrots, onions, soft fruits, clovers.
Single ingredient products	Fusilade, Power Prime.
Permitted mixtures	None significant.
Risks and precautions	Irritating to eyes and skin; harvest intervals: beets and carrots 8 weeks, onions 4 weeks; fruit crops not to be treated after flowering.

Fluroxypyr

Function	Post-emergence herbicide [heterocyclic].
Uses	Against annual dicotyledons in cereals and grassland.
Single ingredient products	Starane.
Permitted mixtures	With bromoxynil, clopyralid, cyanazine, ioxynil.
Products based on mixtures	Advance, Hotspur, Sickle, Spitfire, Stexal.
Risks and precautions	Harmful to livestock; keep livestock out of treated fields for at least 3 days.

Fonofos

Function	Soil- and seed-treating insecticide [organophosphorus].
Uses	Against frit fly, wheat bulb fly, and wireworm in cereals; cabbage stem flea beetle and cabbage root fly in brassicas; sciarid fly and vine weevil in hardy ornamentals and nursery fruit trees or bushes.
Single ingredient products	Cudgel, Dyfonate, Fonofos Seed Treatment.
Permitted mixtures	With disulfoton.
Products based on mixtures	Doubledown.

Risks and precautions	Harmful if swallowed, breathed, or contacts skin (organophosphorus); dangerous to game, wild birds, and animals; dangerous to fish; treated seed not to be used for feed.

Formaldehyde

Function	Fumigant fungicide.
Uses	Against soil-borne fungi in glasshouses.
Single ingredient products	Dyna-Form, Formaldehyde.
Permitted mixtures	With lindane.
Products based on mixtures	Microgen Plus.
Risks and precautions	Toxic if swallowed; harmful if breathed or contacts skin.

Fosamine-ammonium

Function	Contact herbicide.
Uses	Against woody weeds in non-crop areas and in forestry.
Single ingredient products	Krenite.
Permitted mixtures	None significant.

Fosetyl-aluminium

Function	Systemic fungicide [phosphonate].
Uses	Against fungi in apples, broad beans, hops, strawberries, peas, glasshouse ornamentals, and flowers.
Single ingredient products	Alitte.
Permitted mixtures	With captan, thiabendazole.
Risks and precautions	Irritating to skin, eyes, and lungs; harvest intervals: hops 4 days, broad beans 17 days, apples 24 days.

Fuberidazole

Function	Fungicide [heterocyclic, MBCs].
Uses	Against fungi on cereal seeds.
Single ingredient products	Available only in mixtures.
Permitted mixtures	With ethirimol, imazalil, triadimenol, triamenol.
Products based on mixtures	Baytan.

Furalaxyl

Function	Fungicide.
Uses	Against fungi in ornamentals, nursery stock, pot plants, bedding plants.
Single ingredient products	Fongarid.
Permitted mixtures	None significant.
Risks and precautions	Irritating to eyes and skin.

Glyphosate

Function	Systemic herbicide [amides and ureas].
Uses	Against annual and perennial weeds including couch grass in a wide range of food crops, fruit trees, forestry, grass, amenity grass.
Single ingredient products	Mascot Sonic, Muster, Power Glyphosate, Roundup, Spasor, Sting, Stirrup, **Greenscape Ready To Use Weed Killer**, **Murphy Tumbleweed**.
Permitted mixtures	With simazine.
Products based on mixtures	Mascot Ultrasonic, Rival.
Risks and precautions	Harmful if swallowed; irritating to eyes and skin; dangerous to fish.

Guazatine

Function	Seed-dressing fungicide.
Uses	Seed treatments for wheat.
Single ingredient products	Rappor.
Permitted mixtures	With imazalil.
Products based on mixtures	Rappor Plus.
Risks and precautions	Harmful if swallowed; irritating to eyes.

Heptenophos

Function	Contact and systemic fungicide [organophosphorus].
Uses	Against aphids, American serpentine leaf miner, and thrips in cereals, brassicas, cucumbers, marrows, soft fruits, tomatoes, peppers, glasshouse crops, pot plants.
Single ingredient products	Hostaquick.
Permitted mixtures	With permethrin.
Products based on mixtures	**Murphy Tumblebug**.
Risks and precautions	Toxic if swallowed, breathed, contacts skin (organophosphorus); dangerous to bees.

Imazalil

Function	Systemic fungicide (heterocyclic).
Uses	Against powdery mildew in cucumbers, roses, glasshouse ornamentals, and roses.
Single ingredient products	Fungaflor, Fungaflor Smoke.
Permitted mixtures	With fuberizadole, thiabendazole, triamenol.
Products based on mixtures	Seedtect, Baytan Im.

Risks and precautions	Irritating to eyes and skin; harmful to bees; harvest interval: 1 day for cucumbers.

Imazamethabenz-methyl

Function	Post-emergence herbicide.
Uses	Against wild oats, blackgrass, charlock, and rape as a weed in winter wheat and barley.
Single ingredient products	Daggar.
Permitted mixtures	With isoproturon.
Products based on mixtures	Pinnacle.
Risks and precautions	Irritating to eyes.

Imazapyr

Function	Systemic herbicide.
Uses	For total vegetation control in non-crop areas.
Single ingredient products	Arsenal.
Permitted mixtures	With atrazine.
Products based on mixtures	Arsenal XL.
Risks and precautions	Irritating to eyes; not to be used on food crops.

Iodofenphos

Function	Contact and internal insecticide [organophosphorus].
Uses	Against caterpillars in brassicas; onion fly.
Single ingredient products	Elocril, Rentokil Cockroach Bait, Rentokil Waspex, Sorex Crawling Insect Bait.
Permitted mixtures	None significant.
Risks and precautions	Harmful if swallowed, breathed, or contacts skin (organophosphorus); irritating to eyes and skin; dangerous to bees and fish; harvest intervals: Brussels sprouts 7 days, cabbage 14 days.

Ioxynil

Function	Contact herbicide [benzonitrile, HBNs].
Uses	Against annual dicotyledons in onions, turf.
Single ingredient products	Actrilawn, Totril.
Permitted mixtures	With benazolin, bromoxynil, chlorotoluron, chlorsulfuron, clopyralid, 2,4-D, dicamba, dichlorprop, ethofumesate, fluroxypyr, isoproturon, linuron, MCPA, mecoprop, trifluralin.
Products based on mixtures	Actril, Advance, Alpha Briotril, Astrol, Atlas Minerva, Atom, Asset, Assassin, Banvel, Certrol, Certrol-Lin Onions, Crusader, Deloxil, Framolene, Glean, Harrier, Hobane, Hotspur, Iotox, Jaguar, Malet, Masterspray, Mylone, Novacorn, Oxytril, Post Kite, Stellox, Stexal, Super Verdone, Swipe, Terset, Tetrone, Thor.
Risks and precautions	Harmful if contacts skin or is swallowed; irritating to eyes and skin; keep livestock out of treated areas for at least 6 weeks.
Status	Under review.

Iprodione

Function	Fungicide [heterocyclic].
Uses	Against a wide range of fungi on cereals, brassicas, field beans, peas, lettuce, cucumbers, peppers, tomatoes, flowers, nursery fruit trees, ornamentals.
Single ingredient products	Rovral Dust, Tubair Rovral.
Permitted mixtures	With metalaxyl, thiabendazole, thiophanate-methyl.
Products based on mixtures	Polycote Prime, Compass.
Risks and precautions	Irritating to eyes and skin; harvest intervals: strawberries 1 day; tomatoes 1 day; cucumbers 2 days; lettuce, onions, and peppers 7 days; brassicas, rape, and beans 21 days.
Status	Under review.

Isoproturon

Function	Herbicide [amides and ureas].
Uses	Against annual grasses and dicotyledons in cereals.
Single ingredient products	Arelon, Chiltern, Hytane, Portman Isotop, Power Isoproturon, Power Swing, Sabre, Tolkan 500.
Permitted mixtures	With bifenox, diflufenican, isoxaben, mecoprop, metsulfuron-methyl, pendimethalin, tri-allate, trifluralin.
Products based on mixtures	Assassin, Astrol, Autumn Kite, Cougar, Fanfare 469, Foxstar, Hytane Extra, Ipso, Javelin, Musketeer, Oracle, Panther, Pinnacle, Quiver, Terset, Trump.
Risks and precautions	Irritating to skin and eyes.
Status	Under review.

Isoxaben

Function	Soil-acting herbicide [amides and ureas].
Uses	Against annual dicotyledons in cereals, amenity grass, soft fruits.
Single ingredient products	Flexidor, Knot Out, Tripart Ratio.
Permitted mixtures	With isoproturon, methabenzthiazuron.
Products based on mixtures	Fanfare 469, Glytex, Ipso.

Lenacil

Function	Residual, soil-acting herbicide [heterocyclic].
Uses	Against annual dicotyledons and annual grasses in beets, soft fruit, flowers, ornamental nursery stock, spinach, and strawberries.
Single ingredient products	Ashlade Lenacil, Venzar, Visor.
Permitted mixtures	With chloridazon, linuron, phenmedipham.
Products based on mixtures	Advizor, DUK 880, Lanside, Varmit.

Risks and precautions	Irritating to eyes, lungs, and skin.
Status	Review planned.

Lindane

Function	Contact and ingested insecticide [organochlorine].
Uses	Against a wide range of insects in food crops, ornamentals, forestry; industrial and household pests.
Single ingredient products	(Note: this is only a small sample of the large number of available products – check labels carefully.) Antel Woodworm Killer, Atlas Steward, Bio-Chem Dual Purpose, Cube Woodworm Treatment Fluid, FS Lindane Dust, Fumite Lindane, Gamma-Col, Gammasan 30, Hortag Hexaflow Plus, Lindane Flowable, Murfume Grain Store Smoke, Noe Kotol, Protim 90-210, Unicrop Leatherjacket Pellets, Wireworm FS Seed Treatment, **Cuton Ant Powder**, **Doff Ant Killer**, **Gamma BHC Garden Spray**, **Greenhouse Smoke Crawling Pest Killer**, **Grovex Smoke Generator HCH**, **Grovex Smoke Nigets**, **Horticultural Dust**, **Murphy Ant Killer Powder**, **Murphy Gamma BHC Dust**, **Nexa Lotte**, **Soil Insecticides Powder**, **Wilko Ant Killer**, **Woodworm Fluid Concentrate**.
Permitted mixtures	With aldicarb, captan, diazinon, dichlofluanid, dichlorvos, dimethoate, fenitrothion, fenpropimorph, formaldehyde, malathion, phenylmercury acetate, pyrethrins, resmethrin, rotenone, tecnazene, tetramethrin, thiabendazole, thiophanate-methyl, thiourea, thiram, tributylin oxide.
Products based on mixtures	(Note: this is only a small sample of the large number of available products – check labels carefully.) Aquaseal Super, Bio-Chem Fungicide Insecticide, Cube Universal Treatment Fluid, Desanex Dairy Fly Spray, Castaway Plus Gammalex, Fumite Tecnalin Smoke, Grovex Lindane Pyrethrin, Killgerm Lindacide, Killgerm Tetracide Insecticidal Spray, Lindex Plus Seed Treatment, Marstan Flyspray, Mergamma 30,

Microgen Plus, Mystox ATP2, New Hysede RCR Wasp Killer Powder, Rentokil Lindane Pybuthrin Dust, Saturin 5, Sentry, Vitalex, **Ant Killer, Boots Wasp and Fly Killer, Combined Seed Dressing, Doom Moth Proffer Aerosol, Extra Strength Ant Killer, Flora Spray Systemic Insect Killer, Hickson Timber Care, Mole Ban, Murphy Pest and Disease Smoke, PBI Hexyl, Payless Universal Wood Preserver, Preservative For Wood In Four Colours, Pro-Am Timber Treat Green, Rentokil Woodworm Fluid, Secto Rose and Flower Spray, Secto Aphid Killer, Secto Greenfly and Garden Insect Spray, Secto Insect Killer Powder, Secto Wasp & Ant Killer Aerosol, Tecnalin Smoke Cone.**

Risks and precautions	Toxic or harmful if swallowed, breathed, or contacts skin; irritating to eyes and lungs; dangerous to bees; treated seed not to be used for feed; keep all livestock out of treated areas for at least 14 days.
Status	Under review.

Linuron

Function	Contact and residual herbicide [amides and ureas].
Uses	Against a wide range of annual dicotyledons and some grasses in a wide range of cereals, vegetables, and ornamentals.
Single ingredient products	Afalon, Ashlade Linuron, Atlas Linuron, Campbell's Linuron, Defender, Du Pont Linuron, Marksman, Quadrangle Defender, Rotalin.
Permitted mixtures	With bifenox, bromoxynil, chlorpropham, cyanazine, 2,4-DB, ioxynil, lenacil, MCPA, mecoprop, terbutryn, trietazine, trifluralin.
Products based on mixtures	Alistell, Ashlade Flint, Atlas Janus, Broadcide 20, Bronox 20, Campbell's Solo, Chandor, Clovacorn Extra, Certriol-Lin Onions, Janus, Linnet, Marksman Solo, Neminfest, Portman Trilin, Pre-Empt, Profalon, Quadrangle Onslaught, Stay-Kleen, Tempo, Warrior.
Risks and precautions	Irritating to eyes, lungs, and skin; do not graze crops within 4 weeks of treatment.
Status	Under review.

Malathion

Function	Contact insecticide [organophosphorus].
Uses	Against a wide range of insects like aphids, pollen beetles, codling moths, thrips in a wide range of vegetables, fruits, and flowers.
Single ingredient products	Malathion, Rentokil Malathion Dust, Tubair Malathion, **Agrichem Greenfly Spray**, **Ban-Mite**, **Doff Whitefly Killer**, **Duramitex**, **Green Fly Aerosols Spray**, **Murphy Malathion Spray/Liquid**.
Permitted mixtures	With captan, dimethoate, lindane, permethrin, pyrethrins.
Products based on mixtures	Evershield Captan/Malathion, Cromocide Spray Conc, **Bio Crop Saver**.
Risks and precautions	Harmful if swallowed, breathed, or contacts skin (organophosphorus); harmful to bees.
Status	Review planned.

Maleic hydrazide

Function	Growth and sprout suppressant.
Uses	For growth suppression in amenity grass, amenity trees and shrubs; sprout suppression in potatoes and onions.
Single ingredient products	Bos MH, Burtolin, Chiltern Fazor, Mazide 25, Regulox K, Royal Slo-Gro, **Stop Gro G8**.
Permitted mixtures	With chlorpropham, 2,4-D, dicamba, MCPA.
Products based on mixtures	BH Dockmaster, Dock Killer, Synchemicals Mazide Selective.

Mancozeb

Function	Fungicide [carbamate, EBDC].
Uses	Against fungi like septorias, rust, sooty mould, and mildews in cereals, rape, lettuce; blight in potatoes; scab in apples; and black spot in roses.
Single ingredient products	Ashlade Mancozeb, Dithane 945, Karamate, Penncozeb, Portman Mancozeb, Unicrop Mancozeb, **PBI Dithane 945**.

Permitted mixtures	With benalaxyl, carbendazim, cymoxanil, 4-indol-3-ylbutyric acid, maneb, metalaxyl, oxadixyl, prochloraz, zineb.
Products based on mixtures	Ashlade Blight Fungicide, Curzate, Fubol, Fytospore, Galben, Kascade, Kombat, Nova, Recoil, San 537F, Systol, Tubair Dicamate.
Risks and precautions	Irritating to lungs; may cause skin sensitisation; harvest intervals: apples, pears, potatoes 7–28 days; lettuce 21 days; grape vines 30 days; blackcurrants and gooseberries 1 month.
Status	Under review.

Maneb

Function	Fungicide [carbamate, EBDC].
Uses	Against rusts, septoria, sooty moulds, and rhynchosporium, in cereals; alternaria and moulds in rape; blight in tomatoes and potatoes; black spot in roses.
Single ingredient products	Ashlade Maneb Flowable, Campbell's X-Spor SC, Clifton Maneb, Headland Spirit, Maneb 80, Manzate, Trimangol 80, Unicrop Maneb Flowable, Unicrop Maneb.
Permitted mixtures	With carbendazim, chlorothalonil, copper oxychloride, fentin acetate, fentin hydroxide, ferbam, mancozeb, sulphur, tridemorph, zinc, zineb.
Products based on mixtures	Ashlade Maneb Plus, Ashlade MNC, Bolda, Brestan 60, Cleanacres Manex, Cosmic, Hytin, Mazin, Quadrangle Manex, Trimastan, Trimanzone, Tripart Senator, Vassgro Manex.
Risks and precautions	Harmful if swallowed; irritating to skin, lungs, and eyes; may cause sensitisation of skin; harvest intervals: outdoor celery 14 days, outdoor crops 7 days, protected crops 2 days.
Status	Under review.

MCPA

Function	Systemic herbicide [chlorinated phenoxy].

Uses	Against annual and perennial dicotyledons in cereals, grassland, amenity grass, and road verges.
Single ingredient products	Agrichem MCPA, Agricorn, Agritox, Agroxone, Atlas MCPA, BASF MCPA Amine, Campbell's MCPA, Chafer MCPA, Phenoxylene 50, Power MCPA, Quad MCPA, Star MCPA.
Permitted mixtures	With amitrole, asulam, atrazine, benazolin, bentazone, bromoxynil, clopyralid, cyanazine, 2,4-D, 2,4-DB, dalapon, dicamba, dichlorprop, diuron, ferrous sulphate, ioxynil, linuron, maleic hydrazide, MCPB, mecoprop, simazine, 2,3,6-TBA.
Products based on mixtures	Agrichem DB Plus, Alistell, Alphanox, Acumen, Actril, Atlas Minerva, Banlene Plus, Banvel M, Campbell's New Camppex, Campbell's Redlegor, Clovacorn Extra, Dock Ban, Docklene, Embutox Plus, Envoy, Farmon 2,4-D Plus, Fisons Green Master Extra, Fisons Tritox, Hemoxone, Herrisol, Hyprone, Hysward, Lontrel Plus, Marks 2,4-D Extra, Mascot Super Selective, MSS Mircam Plus, Paddox, Pasturol, Quadban, Rasinox R, Selectrol, Seritox 50, Serramix, Springcorn Extra, Synchemicals Showerproof Turf Weedkiller, Tetralex-Plus, Tetrone, Tribute, Tropotox Plus, Vergemaster.
Risks and precautions	Harmful if swallowed, or contacts skin; irritating to eyes; risk of serious damage to eyes.
Status	Review planned.

MCPB

Function	Systemic herbicide [chlorinated phenoxy].
Uses	Against annual and perennial dicotyledons in undersown cereals, grass leys, peas, clover-seed crops, edible podded peas.
Single ingredient products	Campbell's Bellmac Straight, Cropsafe MCPB, Marks MCPB, Tropotox.
Permitted mixtures	With bentazone, MCPA.
Products based on mixtures	Campbell's Bellmac Plus, Farmon MCPB-MCPA, Pulsar, Trifolex-Tra, Tropotox Plus.

Risks and precautions	Harmful if swallowed or contacts skin.
Status	Review planned.

Mecoprop

Function	Systemic herbicide [chlorinated phenoxy].
Uses	Against annual and perennial dicotyledons like chickweed and cleavers in cereals, grass-seed crops, grass, turf.
Single ingredient products	Atlas CMPP, Campbell's CMPP, Chafer CMPP, Clencorn, Clovotox, Compitox Extra, Farmon CMPP, Headland Charge, Herrifex DS, Hymec, Iso-Cornox 57, Mascot Cloverkiller, Methoxone.
Permitted mixtures	With amitrole, asulam, atrazine, benazolin, bentazone, bifenox, bromoxynil, clopyralid, cyanazine, 2,4-D, 2,4-DB, dicamba, dichlorophen, dichlorprop, fenoprop, ferrous sulphate, ioxynil, isoproturon, linuron, MCPA, simazine, 2,3,6-TBA, triclopyr.
Products based on mixtures	Assassin, Agrichem Mortweed Plus, Banlene Plus, Banvel, Broxolon, Campbell's Oxystat, Clenecorn Plus, Cleaval, Com-Trol, Dock Ban, Docklene, Fettel, Foxstar, Harrier, Herrisol, Hyban, Hygrass, Hyprone, Hysward, Iotox, Jaguar, Killgerm Special Turf Weedkiller, Malet, Mascot Selective CDA, Mascot Selective Weedkiller, Musketeer, Nintex, Paddox, Pasturol, Paranox, Post-Kite, Primanox, Select-Trol, Seloxone, Supertox 30, Sydex, Swipe, Tetralex, Thor, Tribute, Vedone CDA, Zennapron, **Asda Garden Lawn Weed and Feed/+ Mosskiller, Bio Weed Pencil, Boots Kill-A-Weed, Boots Lawn Weedkiller, Boots Lawn Weed and Feed, Boots Total Lawn Treatment, Conc Selective Weedkiller, Fisons Lawncare Liquid, Fisons New Evergreen 90, Garden Centre Weedspray For Lawns, Gateway Lawn Feed with weedkiller, Gem Lawn Weed and Feed, Lawn Spot Weed Granules, New Formulation SBK Brushwood Killer, Spot Weeder, Super Green and Weed, Supertox, Verdone 2, Wilko Lawn Feed'N'Weed, Wilko Lawn Food With Weedkiller.**

Risks and precautions	Harmful if swallowed or contacts skin; irritating to eyes and skin.
Status	Review planned.

Mephosfolan

Function	Systemic insecticide [organophosphorus].
Uses	Against damson-hop aphid on hops.
Single ingredient products	Cytro-Lane.
Permitted mixtures	None significant.
Risks and precautions	Toxic if swallowed, breathed, or contacts skin; irritating to skin; dangerous to game, wild birds, and animals, dangerous to bees and fish.

Mercuric oxide

Function	Fungicide paint.
Uses	Against canker in apples and pears.
Single ingredient products	Kankerex.
Permitted mixtures	None significant.
Risks and precautions	Harmful if swallowed.

Mercurous chloride

Function	Soil-applied fungicide.
Uses	Against clubroot in brassicas and white rot in onions.
Single ingredient products	Calomel, Mercurous Chloride.
Permitted mixtures	None significant.
Risks and precautions	Harmful if swallowed.

Metalaxyl

Function	Fungicide.

Uses	Against fungi, including mildews in peas, beans, lettuce.
Single ingredient products	Available only in mixtures.
Permitted mixtures	With captan, thiabendazole, thiram.
Products based on mixtures	Apron Combi 453FS, Favour 600FW.
Risks and precautions	Harmful if swallowed; irritating to eyes.

Metaldehyde

Function	Molluscicide.
Uses	Against slugs and snails.
Single ingredient products	Agrichem Slug Pellets, Chiltern Metaldehyde Slug Killer Mini Pellets, Doff Metaldehyde Slug Killer Mini Pellets, Farmon Mini Slug Pellets, Gastratox Mini Slug Pellets, Helarion, Mifaslug, PBI Slug Pellets, Power Metaldehyde 6P, Quad Mini Slug Pellets, **B&Q Slug Killer Blue Mini Pellets, Boots Slug Destroyer Pellets, Cuton Slug Killer Blue Mini Pellets, Doff Quality Slug Killer Blue Mini Pellets, ICI Slug Pellets, Imp Slug Pellets, Imp Slugtape, Murphy Slugits, PBI Slug Mini Pellets, Scram Slug Killer, Secto Slug-Kil Pellets, Tip Top Slug Pellets, Wilko Slug Killer.**
Permitted mixtures	With thiram.
Products based on mixtures	**Fisons Slug & Snail Killer**.
Risks and precautions	Dangerous to game, wild birds, and animals.

Metamitron

Function	Contact and residual herbicide [heterocyclic].
Uses	Against annual dicotyledons and grasses in beets.
Single ingredient products	Goltix, Power Countdown.
Permitted mixtures	None significant.

Methacrifos

Function	Insecticide and acaricide.
Uses	Against storage pests in grain stores.
Single ingredient products	Damfin 950EC.
Permitted mixtures	None significant.
Risks and precautions	Harmful if swallowed, breathed, or contacts skin (organophosphorus); irritating to eyes and skin; harmful to game, wild birds, and animals.

Metham-sodium

Function	Soil sterilant [carbamate, EBDC].
Uses	Against soil-borne diseases in glasshouse crops, flowers, ornamentals; outdoor crops like potatoes, onions; Dutch elm disease.
Single ingredient products	Campbell's Metham Sodium, Sistan, Super Sistan.
Permitted mixtures	None significant.
Risks and precautions	Harmful if swallowed or contacts skin; irritating to eyes, lungs, and skin.

Methiocarb

Function	Stomach-acting molluscicide and insecticide [carbamate].
Uses	Against slugs and snails in field crops and glasshouses; leatherjackets in cereals; seed beetles in strawberries.
Single ingredient products	Club, Draza.
Permitted mixtures	None significant.
Risks and precautions	Is an anticholinesterase compound and should be avoided by people advised not to work with such chemicals; keep poultry out of treated areas for at least 7 days; harmful to fish.

Methomyl

Function	Insecticide [carbamate].
Uses	Against flies in animal houses.
Single ingredient products	Sorex Golden Fly Bait.
Permitted mixtures	None significant.
Risks and precautions	Is an anticholinesterase compound and should be avoided by people advised not to work with such chemicals; keep baits away from children; harmful to birds; dangerous to fish.

Metoxuron

Function	Contact and residual herbicide [amides and ureas].
Uses	Against annual dicotyledons and grasses in cereals, carrots, parsnips.
Single ingredient products	Dosaflo.
Permitted mixtures	With fentin hydroxide, simazine.
Products based on mixtures	Atlas Hermes, Endspray, Hermes.

Metsulfuron-methyl

Function	Contact and residual herbicide [amides and ureas].
Uses	Against annual dicotyledons like chickweed and mayweed in cereals.
Single ingredient products	Ally.
Permitted mixtures	With isoproturon, thifensulfuron-methyl.
Products based on mixtures	Harmony, Oracle.

Monolinuron

Function	Residual herbicide [amides and ureas].

Uses	Against annual dicotyledons, grasses, polygonums, and fat hen in potatoes, leeks, and french beans.
Single ingredient products	Aressin.
Permitted mixtures	With linuron, paraquat.
Products based on mixtures	Broadcide, Gramonol 50.
Risks and precautions	Irritating to skin, lungs, and eyes.
Status	Under review.

Myclobutanil

Function	Systemic fungicide.
Uses	Against scab and powdery mildew in apples and pears; mildew, black spot in roses.
Single ingredient products	Systhane.
Permitted mixtures	None significant.

Nabam

Function	Soil fungicide [carbamate, EBDC].
Uses	Against root rot in tomatoes and phoma rot in chrysanthemums.
Single ingredient products	Campbell's Nabam Soil Fungicide.
Permitted mixtures	None significant.
Risks and precautions	Harmful if swallowed; irritating to eyes, lungs, and skin.
Status	Under review.

1-naphthylacetic acid

Function	Plant growth regulator and rooting hormone.
Uses	For preventing harvest drop in apples; sucker inhibition in tree fruits; rooting of cuttings.

Single ingredient products	Phyomone, Rhizopon, Tipoff.
Permitted mixtures	With captan, dichlorophen, 4-indole-3-ylbutyric acid, thiophanate-methyl, thiram.
Products based on mixtures	Rooting Gel, Synergol, **Bio Roota**, **Boots Hormone Rooting Powder**, **Corry's Rooting Powder**, **Doff Hormone Rooting Powder**, **Green Fingers Hormone Rooting Powder**, **Keri-Root**, **Murphy Rooting Hormone**.

Nicotine

Function	Contact insecticide.
Uses	Against a wide range of insects like aphids, capsids, leaf miners, sawflies, woolly aphids, thrips in a range of crops including vegetables, fruits, flowers, glasshouse crops, and mushrooms.
Single ingredient products	Campbell's Nico Soap, Nicotine 40% Shreds, XL All Insecticide.
Permitted mixtures	None significant.
Risks and precautions	Harmful if swallowed, breathed, or contacts skin; dangerous to game, wild birds, and animals; harmful to bees; dangerous to fish; keep all livestock out of treated area for at least 12 hours.
Status	Review planned.

Nuarimol

Function	Systemic fungicide.
Uses	Against powdery mildew, septoria, and rhynchosporium in cereals.
Single ingredient products	Chemtech Nuarimol, Triminol.
Permitted mixtures	With captan.
Products based on mixtures	Kapitol.
Risks and precautions	Irritating to skin and eyes.

Omethoate

Function	Systemic insecticide and acaricide [organophosphorus].
Uses	Against frit and yellow cereal flies in cereals; aphids in plums and hops.
Single ingredient products	Folimat.
Permitted mixtures	None significant.
Risks and precautions	Toxic if swallowed, breathed, or contacts skin (organophosphorus); harmful to game, wild birds, and animals; harmful to bees; harmful to livestock; keep all livestock out of treated areas for at least 7 days; harvest intervals: cereals 3 weeks, plums 5 weeks, grass crops 6 weeks.

Oxadiazon

Function	Residual herbicide.
Uses	Against annual dicotyledons and grasses in container-grown ornamentals, fruit trees, soft fruits, hops.
Single ingredient products	Ronstar.
Permitted mixtures	None significant.
Risks and precautions	Irritating to eyes; dangerous to fish.

Oxydemeton-methyl

Function	Contact, fumigant, and systemic insecticide [organophosphorus].
Uses	Against a wide range of insects like aphids, pea midges, leafhoppers, red spider mites in cereals, vegetables, fruit trees, soft fruits, flowers, ornamentals.
Single ingredient products	Metasystox.
Permitted mixtures	None significant.

Risks and precautions	Toxic if swallowed, breathed, or contacts skin; harmful to game, wild birds, and animals; harmful to bees; keep all livestock out of treated fields for at least 7 days; leys must not be grazed for 6 weeks.
Status	Under review.

Paclobutrazol

Function	Plant growth regulator.
Uses	For stem shortening in ornamentals and flowers; improving colour in poinsettias; lodging control in cereal seed crops.
Single ingredient products	Bonzi, Cultar, Parlay.
Permitted mixtures	None significant.
Risks and precautions	Do not use on food crops; keep livestock out of treated areas for at least 6 weeks; harvest interval: apples and pears 6 weeks.

Paraquat

Function	Non-selective herbicide [heterocyclic].
Uses	Against dicotyledons and annual grasses in cereals, field crops, vegetables, ornamentals, potatoes, beets, orchards, soft fruits, forestry, non-crop areas.
Single ingredient products	Dextrone, FAL Paraquat, Gramoxone, Power Paraquat, Scythe, Speedway.
Permitted mixtures	With amitrole, bromacil, diquat, diuron, monolinuron, simazine.
Products based on mixtures	Cleansweep, Dexuron, Farmon PDQ, Gramonol, Groundhog, Parable, Soltair, **Pathclear**, **Weedol**.
Risks and precautions	Toxic if swallowed – paraquat can kill; harmful in contact with skin; irritating to eyes and skin; do not put in a drinks bottle; not for use by amateur gardeners; may be harmful to hares; harmful to animals; keep all livestock out of treated areas for at least 24 hours.
Status	Review planned.

Pendimethalin

Function	Residual herbicide [amides and ureas].
Uses	Against annual grasses and dicotyledons like cleavers and speedwell in cereals, potatoes, tree fruits, soft fruits.
Single ingredient products	Stomp.
Permitted mixtures	With isoproturon.
Products based on mixtures	Trump.
Risks and precautions	Irritating to skin and eyes; dangerous to fish.

Pentachlorophenol

Function	Herbicide, fungicide, and insecticide [chlorinated phenoxy].
Uses	As wood protectant, pre-harvest defoliant, and general herbicide.
Single ingredient products	Mystox LP, Cementome Exterior Wood Preservative, Protim Exterior Brown, **Gainserv**, **Sadolins Joinery Preservative**.
Permitted mixtures	With boric acid, bromacil, lindane, tributyltin oxide, zinc naphthenate.
Products based on mixtures	Celpruf PK WR, Cuprinol Wood Preserver Clear, Fenocil, Lowes Volatile Deodex, Pro-Am Timber Treat GP, Protim, Protim Joinery Lining, Rentokil Joinery Injection Fluid, **Rentokil Preservative for Wood Clear**, **Rentokil Supergrade Wood Preserver**, **Signpost Wood Preservative Clear**, **Wickes Clear Wood Preserver**, **Wilko Clear Wood Preserver**.
Risks and precautions	Harmful if swallowed or contacts skin; irritating to skin, lungs, and eyes; keep children and unprotected persons off treated ground for at least 2 weeks or until after heavy rain.
Status	Review planned.

Pentanochlor

Function	Contact herbicide [amides and ureas].
Uses	Against annual dicotyledons, annual meadow grass in vegetables, soft fruits, flowers, nursery stock.
Single ingredient products	Atlas Solan, Croptex Bronze.
Permitted mixtures	With chlorpropham.
Products based on mixtures	Atlas Brown.
Risks and precautions	Irritating to eyes, lungs, and skin.
Status	Review planned.

Permethrin

Function	Contact and ingested insecticide [pyrethroid].
Uses	Against a wide range of insects like aphids, codling moths, whitefly, leaf miners, in apples, pears, vegetables, flowers, and grain stores.
Single ingredient products	Cooper Coopex Smokes, Darmycel Agarifume Smoke, Fumite Permethrin Smoke, Permasect, Tubair Permethrin, Rentokil Permethrin Wettable Powder, Turbair Permethrin, Young's Blitsect 25, **Baby Bio House Plant Insecticide, Bio Flydown, Bio One-Shot For Insect Pests, Bio Sprayday, Boots Caterpillar and Whitefly Killer, Darmycel Agarifume Smoke Generator, Fisons Insect Spray For Houseplants, Fumits Whitefly Smoke Cone, Murphy Whitefly Smoke, Murphy Zap Cap General Insecticide, Picket, Rapitest Insect Spray, Secto Extra Strength Insect Killer, Synchemicals Houseplant Leaf Shine & Plant Pest Killer, Tesco House Plant Leaf Shine.**
Permitted mixtures	With bioallethrin, carbendazim, copper oxychloride, dichlofluanid, dichlorvos, dimethoate, dinocap, fenarimol, heptenophos, malathion, pirimiphos-methyl, sulphur, tetramethrin, thiabendazole, triforine.

Products based on mixtures	Combinex, Imperator, Residroid, **Bio Crop Saver**, **Bio Long Last**, **Bio Multirose**, **Bio Multiveg**, **Doom Fly & Wasp Killer**, **Dragon**, **Floracid**, **Kill-a-Bug**, **Murphy Systemic Action Insecticide**, **Mystic General Garden Insecticide**, **Nippon Ant & Crawling Insect Killer**, **Scram Fly Killer**, **Selkill**, **Sybol Aerosol**.
Risks and precautions	Irritating to eyes, lungs, and skin; dangerous to bees; extremely dangerous to fish.
Status	Under review.

Phenmedipham

Function	Contact herbicide [carbamate].
Uses	Against annual dicotyledons in beets and strawberries.
Single ingredient products	Atlas Protrum, Beetomax, Betanal, Campbell's Beetup, Goliath, Gusto, Headland Dephend, Pistol 400, Portman Betalion, Power Phenmedipham, Suplex, Tripart Beta, Vanguard.
Permitted mixtures	With clopyralid, ethofumesate.
Risks and precautions	Harmful if swallowed, breathed, or contacts skin; irritating to eyes, lungs, and skin.

Phenothrin

Function	Contact and ingested insecticide [pyrethroid].
Uses	Against flies in livestock houses.
Single ingredient products	Sumithrin.
Permitted mixtures	With tetramethrin.
Products based on mixtures	Killgerm ULV 500, Sorex Super Fly Spray.
Risks and precautions	Irritating to eyes; dangerous to fish.

Phorate

Function	Systemic insecticide [organophosphorus].

Uses	Against aphids, capsids, carrot fly, bean and pea weevils, wireworms, and other insects in brassicas, beets, beans and peas, potatoes, carrots, lettuce.
Single ingredient products	BASF Phorate, Campbell's Phorate, Chafer Phorate Granules, Mandops Phorate 10, Terrathion Granules.
Permitted mixtures	None significant.
Risks and precautions	Toxic if swallowed, breathed, or contacts skin (organophosphorus); dangerous to game, wild birds, and animals; dangerous to fish; keep all livestock out of treated areas for at least 6 weeks.
Status	Review planned.

Picloram

Function	Translocated herbicide [heterocyclic].
Uses	Against annual and perennial dicotyledons on non-crop areas.
Single ingredient products	Tordon 22K.
Permitted mixtures	With bromacil, 2,4-D.
Products based on mixtures	Atladox, Hydon, Tordon 101.
Risks and precautions	Irritating to eyes.
Status	Review planned.

Pirimicarb

Function	Insecticide [carbamate].
Uses	Against aphids in a wide range of field crops, fruit trees and bushes, flowers, and ornamentals.
Single ingredient products	Aphox, Phantom, Pirimicarb 50DG, Pirimor, Power Pursuit, Sapir, **Mystic Greenfly Killer**, **Rapid Aerosol**, **Rapid Greenfly Killer**.
Permitted mixtures	With bupirimate, triforine.
Products based on mixtures	**Roseclear**.

Risks and precautions	Harmful if swallowed (anticholinesterase compound); harmful to livestock; keep all livestock out of treated areas for at least 7 days; harvest intervals: cereals, lettuce under glass 14 days; cucumbers, tomatoes, and peppers under glass 2 days.

Pirimiphos-ethyl

Function	Insecticide [organophosphorus].
Uses	Against phorid and sciarid flies in mushrooms, bedding and pot plants.
Single ingredient products	Fernex.
Permitted mixtures	None significant.
Risks and precautions	Harmful if swallowed (organophosphorus); dangerous to game, wild birds, and animals; dangerous to fish.

Pirimiphos-methyl

Function	Contact and fumigant insecticide [organophosphorus].
Uses	Against a wide range of insects like wheat bulb fly, frit fly in cereals; weevils in stored grains; cabbage stem flea beetles in brassicas; carrot fly, apple sawfly, codling moths in apples and pears; thrips, whitefly in glasshouses.
Single ingredient products	Actellic, Blex, **Ant Killer Dust**, **Fumite General Purpose Insecticide Smoke Cone**, **Soil Pests Killer**, **Sybol 2**, **Tesco Antkiller**, **Tesco Garden Insecticide**.
Permitted mixtures	With permethrin, pyrethrins.
Products based on mixtures	Kerispray, Sybol 2 Aerosol.
Risks and precautions	Organophosphorus compound; irritating to eyes and skin; ventilate fumigated areas thoroughly before re-entry.

Prochloraz

Function	Broad-spectrum fungicide [heterocyclic].
Uses	Against fungi like eyespot, glume blotch, rhynchosporium in cereals; alternaria, grey mould, and stem rot in rape; fungal diseases of ornamentals and grass-seed crops.
Single ingredient products	Octave, Power Prochloraz, Sporgon, Sportak.
Permitted mixtures	With carbendazim, fenpropimorph, mancozeb, manganese.
Products based on mixtures	Nova, Sporgon WP, Sportak Alpha, Sprint.
Risks and precautions	Harmful if swallowed, breathed, or contacts skin.

Prometryn

Function	Contact and residual herbicide [heterocyclic, triazine].
Uses	Against annual dicotyledons and grasses in peas, potatoes, carrots, parsley, leeks.
Single ingredient products	Atlas Prometryne, Gesagard.
Permitted mixtures	With terbutryn.
Products based on mixtures	Peaweed.
Risks and precaut˙ ˙ns	Harvest interval: 6 weeks.

Propachlor

Function	Pre-emergence herbicide [amides and ureas].
Uses	Against annual dicotyledons and grasses in brassicas, swedes, strawberries, ornamentals.
Single ingredient products	Albrass, Atlas Orange, Croptex Amber, Portman Propachlor, Ramrod Flowable, Tripart Sentinel, **Murphy Covershield Weed Preventer**, **No-Weed**.
Permitted mixtures	With chloridazon, chlorthal dimethyl.

Products based on mixtures	Ashlade CP, Decimate.
Risks and precautions	Irritating to eyes, lungs, and skin.

Propham

Function	Pre-emergence herbicide [carbamate].
Uses	Against annual grasses, chickweed, polygonums in beets and peas.
Single ingredient products	MSS IPC 50 WP, Triherbicide, **Tumbleblite**.
Permitted mixtures	With chloridazon, chlorpropham, diuron, fenuron.
Products based on mixtures	Atlas Electrum, Atlas Gold, Atlas Indigo, Atlas Pink, Barrier, Campbell's Sugarbeet Herbicide, Marks PCF Beet Herbicide, Pommetrol, Power Gro-Stop, Profen, Truchem Quintex.

Propiconazole

Function	Systemic fungicide [heterocyclic, triazole].
Uses	Against fungal diseases in cereals, grass-seed crops, rape, and chrysanthemums.
Single ingredient products	Power Propiconazole, Radar, Tilt.
Permitted mixtures	With carbendazim, chlorothalonil, fenpropimorph, tridemorph.
Products based on mixtures	Glint, Hispor, Sheen, Sambarin, Tilt Turbo.
Risks and precautions	Irritating to eyes, lungs, and skin; dangerous to fish; harvest intervals: cereals 5 weeks, rape 4 weeks.

Propyzamide

Function	Residual herbicide [amides and ureas].
Uses	Against a wide range of annual dicotyledons and annual and perennial grasses in non-cereal field

crops, fruit trees, woodlands, herbs, ornamentals.

Single ingredient products Campbell's Rapier, Kerb Flo, Kerb Flowable, Power Propyzamide, Rapier.

Permitted mixtures With clopyralid.

Products based on mixtures Matrikerb.

Risks and precautions Harvest interval: 6 weeks.

Pyrazophos

Function Systemic fungicide and insecticide [organophosphorus].

Uses Against fungi in cereals, Brussels sprouts, apples, hops, roses, and pot plants; American serpentine leaf miners, western flower thrips in ornamentals, beet-root, leaf brassicas, carrots, cucumbers, celeriac.

Single ingredient products Afugan, Missile, **Pokon Mildew Spray**.

Permitted mixtures None significant.

Risks and precautions Harmful if swallowed (organophosphorus); irritating to eyes, lungs, and skin; harmful to game, wild birds, and animals; dangerous to bees and fish; harvest interval: 2 weeks.

Pyrethrins

Function Contact insecticide.

Uses Against weevils in stored grain, flies. Used as household and garden insecticide.

Single ingredient products Alfadex, Killgerm, Pyra-fog, **Aquablast Bug Spray**, **Big D Fly & Wasp Killer**, **Bug Gun**, **Cromessol Flying Insect Killer**, **Cuton Fruit and Vegetable Insecticide Spray**, **Cuton Rose and Flower Insecticide**, **Doff Fruit and Vegetable Insecticide**, **Fine Fare Pestspray for Fruit and Vegetables**, **Greenfly Spray**, **PBI Anti-Ant Duster**, **Py Garden Insecticide**, **Py Powder**, **Pyrethrum Garden Insect Killer**, **Secto Fly &**

Wasp Killer, Sherley's Vamoose All-Purpose Insecticide, Spraydex Garden Insect Killer, Winfield Flykiller.

Permitted mixtures	With *Bacillus thuringiensis*, bendiocarb, carbaryl, diazinon, dichlorvos, lindane, malathion, pirimiphos-methyl, resmethrin, rotenone.
Products based on mixtures	Cromocide Spray Conc, Ficam Plus, Grovex Lindane Pyrethrin, Killgerm Lindacide, Marstan Flyspray, Pynosect 30RCR Wasp Killer Powder, Rentokil Diazinon Oil Spray, Rentokil Lindane Pybuthrin Dust, Rodesco Insect Powder No. 4, **Ant Killer, Bactospeine Garden, Bio Back to Nature Spray, Doom Garden Insect Killer Aerosol, Doom House Plant Insect Killer, Doom Moth Proofer Aerosol, Extra Strength Ant Killer, Kerispray, Nuvan Fly Killer, Rentokil Blackfly & Garden Insect Killer, Rentokil Garden & Greenhouse Garden Insect Killer, Secto Derris Dust, Secto Greenfly & Garden Insect Spray, Secto House & Garden Powder, Secto Wasp Killer Powder, Tesco House Plant Pest Killer.**
Risks and precautions	Irritating to eyes, lungs, and skin; do not allow spray to contact food products.

Resmethrin

Function	Contact insecticide [pyrethroid].
Uses	Against aphids and caterpillars in brassicas, beans, soft fruit, cucumbers, tomatoes, mushrooms, flowers, pot plants.
Single ingredient products	Turbair Resmethrin Extra, **Boots House Plant Insecticide.**
Permitted mixtures	With bromophos, fenitrothion, lindane, permethrin, pyrethrins, tetramethrin.
Products based on mixtures	Delsanex Dairy Fly Spray, Killgerm & Marstan Fly Spray Aerosols, Turbair Kilsect Long Life Grade, Turbair Grain Store Insecticide, Wasp Nest Destroyer, **Doom Garden Insect Killer Aerosol, Doom Household Insect Killer, Erin Household Insect Killer, Finsect, Pynosect 40, Rentokil Blackfly & Greenfly Killer, Rentokil**

House Plant Insect Killer, Synchemicals House Plant Pest Killer, Tesco House Plant Pest Killer.

Risks and precautions: Irritating to eyes, lungs, and skin; harmful to bees.

Rotenone

Function: Contact insecticide.

Uses: Against aphids in fruit, vegetables, flowers, ornamentals; raspberry beetles in soft fruit; sawflies in gooseberries.

Single ingredient products: FS Derris Dust, FS Liquid Derris, **Derris Dust, Doff Derris Dust, Murphy Derris Dust, PBI Liquid Derris, Tesco Derris Dust, Wasp Exterminator.**

Permitted mixtures: With carbaryl, lindane, pyrethrins, quassia, sulphur, thiram.

Products based on mixtures: RCR Wasp Killer Powder, **Boots Garden Insect Powder With Derris, Bio Back to Nature Spray, Bio Back to Nature & Disease Duster, PBI Hexel, Secto Derris Dust.**

Risks and precautions: Dangerous to fish; harvest interval: 1 day.

Simazine

Function: Soil-acting herbicide [heterocyclic, triazine].

Uses: Against annual dicotyledons and grasses in beans, fruit trees, soft fruits, asparagus, rhubarb, maize, ornamentals, non-crop areas.

Single ingredient products: Ashlade Simazine, Boroflow, Gesatop, Mascot Simazine Granular, MSS Simazine, Imapron, Simflow, Syngran, Truchem Simazine, Weedex S2 FG, **Weedex.**

Permitted mixtures: With amitrole; 2,4-D; 2,4-DES; diquat; diuron; glyphosate; MCPA; mecoprop; metoxuron; paraquat; trietazine.

Products based on mixtures: Alcatraz, Alpha Simazol, Alphanox, Arrow Total Weedkiller, Atlas Hebron Blue, Atlas Total S, Aventox, Boroflow S/ATA, CDA Simflow Plus, Clearway, Deltanox, Fisons Herbazin Plus, Groundhog, Hermes, Hytrol, Mascot Highway

Liquid, Orchard Hytrol, Primatol SE, Simflow
Plus, Rival, Soltair, TWK, Total Weedkiller,
Transformer Primatol SE, Ultra-sonic, Unicrop
Simatrole, **Finefare Total Weedkiller**, **Fisons
Path Weeds Killer**, **Hytrol**, **Super Weedex**,
Tesco Total Weedkiller Liquid, **Total Weed**,
Total Weedkiller Granules.

Risks and precautions	Irritating to eyes, lungs, and skin.
Status	Review planned.

Sodium chlorate

Function	Non-selective herbicide.
Uses	For total vegetation control in non-crop areas, paths, around buildings.
Single ingredient products	Arpal NonSelex Powder, Atlacide Soluble Powder, Centex, Granular Weedkiller, Lever Sodium Chlorate, **Betterware Path and Drive Deweeder**, **Cooke's Liquid Sodium Chlorate Weedkiller**, **Doff Sodium Chlorate Weedkiller**, **Murphy Sodium Weedkiller**, **Sodium Chlorate**.
Permitted mixtures	With atrazine, 2,4-D.
Products based on mixtures	Atlacide Extra Dusting Powder.
Risks and precautions	Harmful if swallowed; oxidising substance – sprayed material easily flammable.
Status	Review planned.

Sulphur

Function	Broad-spectrum fungicide.
Uses	Against fungal diseases, including mildews, and scab in cereals, fruit trees, soft fruits, ornamentals.
Single ingredient products	Ashlade Sulphur, Chiltern Super Six, Clifton Sulphur, Hortag Aquasulf, Kumulus S, Microsul Flowable Sulphur, Solfa, Thiovit, Tripart Imber, Vassgro Flowable Sulphur, **Green Sulphur**, **Murphy Mole Smokes**, **Phostrogen Safer's Ready-to-Use Garden Fungicide**, **Safer's**

Natural Garden Fungicide, **Sulphur Candles**, **Yellow Sulphur**.

Permitted mixtures	With carbendazim, copper oxychloride, dinocap, maneb, permethrin, rotenone, thiabendazole, triforine.
Products based on mixtures	Ashlade SMC, Ashlade TCNB, Bolda, Tripart Senator, **Bio Multiveg**, **Bio Multirose**, **Bio Back To Nature Pest & Disease Duster**.

Tar Oils

Function	Fungicide and insecticide wash.
Uses	Against aphids, scale insects, moths in fruit trees, grape vines, hops.
Single ingredient products	Mortegg Emulsion, Sterilite Miscible Hop Defoliant, Sterilite Tar Oil Winter Wash 60% Stock Emulsion, **Jeyes Fluid**, **Tar Oil Winter Wash**.
Permitted mixtures	With phenols.
Products based on mixtures	**Murphy Mortegg**.
Risks and precautions	Harmful if swallowed; irritating to eyes, lungs, and skin.
Status	Review planned.

TCA (Trichloroacetic acid)

Function	Soil-applied herbicide.
Uses	Against annual grasses and volunteer cereals in rape, beets, peas, field beans, carrots, brassicas, swedes, turnips.
Single ingredient products	Farmon TCA, MSS TCA, NATA.
Permitted mixtures	None significant.
Risks and precautions	Harmful if swallowed; irritating to eyes, lungs, and skin.
Status	Review planned.

Tecnazene

Function	Fungicide and potato-sprout suppressant.
Uses	For sprout suppression in stored potatoes; botrytis in lettuce, tomatoes, chrysanthemums, ornamentals.
Single ingredient products	Ashlade TCN, Atlas Tecgran, Atlas Tecnazene Dust, Bygram, Fumite TCNB Smoke, Fusarex Dust/Granules, Headland Suppress, Hickstor Granules, Hortag Tecnazene Dust, Hystore, Hytec, Nebulin, Power Tecnazene, Quad-Keep, Tripart Arena, Tubodust, Tubostore Granules, **Greenhouse Smoke Disease Killer**.
Permitted mixtures	With carbendazim, lindane, thiabendazole, thiourea.
Products based on mixtures	Arena Plus, Bygran, DBK Tecnazene/Carbendazim Dust, Fumite Tecnalin Smoke, Headland Suppress, Hickstor 6 Plus MBC, Storaid Dust, Storite SS, Tecnacarb, Tripart Arena TBZ6, **New Improved Murphy's Mole Smoke**, **Mole Ban**.
Risks and precautions	Irritating to eyes and lungs; harvest intervals: tomatoes 2 days, potatoes 6 weeks.
Status	Under review.

Terbuthylazine

Function	Herbicide [heterocyclic, triazine].
Uses	Against annual dicotyledons and grasses in peas, beans, and potatoes.
Single ingredient products	Available only in mixtures.
Permitted mixtures	With atrazine, terbutryn.
Products based on mixtures	Gardoprim, Opogard, Transformer Gardoprim.
Risks and precautions	Harmful if swallowed.

Terbutryn

Function	Residual herbicide [heterocyclic, triazine].
Uses	Against annual dicotyledons and annual grasses in cereals.
Single ingredient products	Clarosan, Prebane 500 FW.
Permitted mixtures	With linuron, prometryn, terbuthylazine, trietazine, trifluralin.
Products based on mixtures	Alpha Terbalin, Ashlade Summit, Lextra, Opogard 500FW, Peaweed, Senate, Tempo, Tripart Laurel, Tripart Opera.

Tetramethrin

Function	Contact insecticide [pyrethroid].
Uses	Against flies in animal houses.
Single ingredient products	**Genpest Insecticidal Aerosol, Premiere Fly & Wasp Killer**.
Permitted mixtures	With diazinon, dichlorvos, fenitrothion, lindane, permethrin, phenothrin, resmethrin.
Products based on mixtures	Brimpex, Deslanex Dairy Fly Spray, Flytak, Grovex, Killgerm ULV 500, Py Kill 25 Plus, Resdrin, Residroid Dairy Fly Spray, Sorex Super Fly Killer, Supa Swat Aerosol, **Crown Crest Fly Spray, Doom Fly & Wasp Killer, Dragon Aerosol, Keen Flying Insect Killer, Keen Flying Insect Killer Faster Knockdown, Keen Superkill Ant & Roach Exterminator, Nippon Ant & Crawling Insect Killer, Pesguard, Rentokil Insectrol Aerosol, Safeway Flyspray, Scram Fly Killer, Secto Aphid Killer, Selkill**.
Risks and precautions	Harmful to fish.

Thiabendazole

Function	Systemic fungicide [heterocyclic, MBCs].
Uses	Against dry rot, silver scurf, skin spot on potatoes; canker in rape; fusarium rots in

grasses; and other fungal diseases in mushrooms, ornamental bulbs, and asparagus.

Single ingredient products	Ceratotect, Hymush, Storite Clear Liquid, Storite Flowable, Tecto Dust, **Murphy Zap Cap Systemic Fungicide**.
Permitted mixtures	With benzalkonium chloride, bromophos, captan, carboxin, cufraneb, ethirimol, flutriafol, fosetyl-aluminium, 8-hydroxyquinoline, imazalil, lindane, metalaxyl, permethrin, sulphur, tecnazene, thiram.
Products based on mixtures	Aliette Extra, Apron, Apron Combi, Arena Plus, Bromotex CSD, Byatran, Bygran F, Fumilite Tecnalin Smoke, Cerevax Extra, Ferrax, Headland Suppress Plus, Hichstor 6 Plus MBC, Hortag Carbotec, Ceratotect, Hybulb, Hymush, Hytec Super, Nemacin, Storaid Dust, Storite SS, Tecnazene Dust KT, Testo Flowable, Tog II, Tripart Arena TBZ6, Tecnacarb, Tubazole, Vincit L, **Murphy (Zap Cap) Vegetable Care**.
Risks and precautions	Harvest interval: potatoes 21 days.
Status	Under review.

Thiofanox

Function	Systemic insecticide [carbamate].
Uses	Against aphids, capsids, leafhoppers in potatoes and beets.
Single ingredient products	Dacamox.
Permitted mixtures	None significant.
Risks and precautions	Harmful if swallowed, breathed, or contacts skin (anticholinesterase); dangerous to game, wild birds, and animals; dangerous to fish.

Thiophanate-methyl

Function	Systemic fungicide [heterocyclic, MBCs].
Uses	Against botrytis, eyespot, fusarium blight, and sooty moulds in cereals, brassicas, rape, field beans, fruit trees, cucumbers.

Single ingredient products	Cercobin Liquid, Mildothane Liquid, Mildothane Turf Liquid, **Fungus Fighter**, **Liquid Club Root Control**, **Seal & Heal**.
Permitted mixtures	With dichlorophen, imazalil, iprodione, lindane, 1-naphthylacetic acid, thiram.
Products based on mixtures	Castaway Plus, Compass, Mildothane Liquid, Rooting Gel.
Status	Under review.

Thiram

Function	Fungicide and animal repellent [carbamate].
Uses	Against damping off in field beans, brassicas, and other field crops; botrytis in strawberries, lettuce, pot plants, flowers and tomatoes.
Single ingredient products	Agrichem Flowable Thiram, Hortag Thiram, Thianosan, Tripomol, Unicrop Thiram.
Permitted mixtures	With carbendazim, carboxin, diazinon, dicloran, fenpropimorph, 4-indole-3-ylbutyric acid, iprodione, lindane, metalaxal, metaldehyde, 1-naphthylacetic acid, permethrin, rotenone, thiabendazole, thiophanate-methyl.
Products based on mixtures	Apron Combi, Combinex, Dal CT, Favour, Lindex FS Seed Treatment, New Hydraguard, New Hysede, Turbair Botryicide, **Boots Hormone Rooting Powder**, **Fisons Slug & Snail Killer**, **Flora Spray Systemic Insect Killer**, **Green Fingers Hormone Rooting Powder**, **PBI Hexyl**, **Secto Garden Powder**, **Secto Rose and Flower Spray**.
Risks and precautions	Irritating to eyes, lungs, and skin.
Status	Review planned.

Triadimefon

Function	Systemic fungicide [heterocyclic, triazole].
Uses	Against rhynchosporium, powdery mildew, rust in cereals, beets, brassicas, apples, hops, soft fruits.

Single ingredient products	Bayleton, Portman Triadimefon.
Permitted mixtures	With carbendazim.
Products based on mixtures	Bayleton BM.
Risks and precautions	Harmful to all livestock; keep livestock out of treated areas for at least 21 days; harvest interval: most fruit and vegetables, 14 days.

Tri-allate

Function	Soil-acting herbicide [carbamate].
Uses	Against wild oats and other annual grasses in cereals, brassicas, carrots, leeks, legumes, field and other beans, mange-tout peas.
Single ingredient products	Avadex BW.
Permitted mixtures	With chlorotoluron, isoproturon.
Products based on mixtures	MON 7942 Granular.
Risks and precautions	Harmful if swallowed or in contact with skin; irritating to eyes and skin.
Status	Review planned.

Triamenol

Function	Systemic fungicide [heterocyclic, triazole].
Uses	Against powdery mildew, rust, rhynchosporium in cereals, beets, turnips, swedes, brassicas.
Single ingredient products	Bayfidan, Spinnaker.
Permitted mixtures	With fuberidazole, imazalil, perchloraz, tridemorph.
Products based on mixtures	Bay UK, Baytan, Dorin.
Risks and precautions	Harmful if swallowed; irritating to eyes; harvest interval: beets and brassicas 14 days.

Triazophos

Function	Contact and ingested insecticide [organophosphorus].
Uses	Against a wide range of insects in a wide range of field crops.
Single ingredient products	Hostathion.
Permitted mixtures	None significant.
Risks and precautions	An organophosphorus compound – not to be used by people medically advised to avoid such compounds; toxic if swallowed; harmful if swallowed or contacts skin; harmful to game, wild birds, and animals; dangerous to bees; harvest interval: 4 weeks.

Trichlorfon

Function	Contact and ingested insecticide [organophosphorus].
Uses	Against flea beetles in beets, caterpillars, and other insects in brassicas, vegetables, apples, soft fruits, tomatoes, strawberries, and roses.
Single ingredient products	Dipterex 80.
Permitted mixtures	None significant.
Risks and precautions	Harvest interval: 14 days.

Triclopyr

Function	Herbicide [heterocyclic].
Uses	Against perennial dicotyledons, woody weeds, dock, nettles, broom, gorse in grassland, forestry, non-crop areas.
Single ingredient products	Garlon, Timbrel, **Murphy Kil-Nettle**.
Permitted mixtures	With clopyralid, 2,4-D, dicamba, mecoprop.
Products based on mixtures	Broadshot, Fettel, Grazon 90.

Risks and precautions	Harmful if swallowed; irritating to skin; dangerous to fish.

Trifluralin

Function	Herbicide [heterocyclic, triazine].
Uses	Against annual grasses and dicotyledons in brassicas, vegetables, soft fruits, parsley and other herbs, ornamentals.
Single ingredient products	Ashlade Trimaran, Atlas Trifluralin, Treflan, Trigard, Tristar.
Permitted mixtures	With bromoxynil, diflufenican, ioxynil, isoproturon, linuron, napropamide, terbutryn, trietazine.
Products based on mixtures	Ardent, Autumn Kite, Ashlade Flint, Atlas Janus, Campbell's Solo, Chandor, Devrinol, Ipicombi TL, Janus, Lextra, Linnet, Marksman Solo, Neminfest, Portman Trilin, Quadrangle Onslaught, Tri-farmon FL, Warrior.
Risks and precautions	Irritating to eyes, lungs, and skin.

Triforine

Function	Systemic fungicide.
Uses	Against powdery mildew in barley; fairy rings in turf; American gooseberry mildew in gooseberries; mildews in ornamentals, flowers, and cucumbers.
Single ingredient products	Fairy Ring Destroyer, Funginex, Saprol, **Celamerck Unit Dose Garden Fungicide**, **Hozelock Re-Cap Fungicide**.
Permitted mixtures	With bupirimate, carbendazim, dinocap, permethrin, pirimicarb, sulphur.
Products based on mixtures	Brolly, Nimrod, **Bio Multirose**, **Nimrod T**, **Roseclear**.
Risks and precautions	Harmful in contact with skin; irritating to eyes.

Zineb

Function	Fungicide [carbamate, EBDC].
Uses	Against blight in potatoes, tomatoes; mildews in lettuce, onions, spinach; rust in carnations, chrysanthemums; other fungal diseases in brassicas.
Single ingredient products	Hortag Zineb Wettable, Tritoftorol, Unicrop Zineb.
Permitted mixtures	With ferbram, mancozeb, maneb.
Products based on mixtures	Trimanzone, Trithac, Turbair Dicamate.
Risks and precautions	Irritating to eyes, lungs, and skin; harvest intervals: soft fruit 4 weeks, lettuce 3 weeks.
Status	Under review.

Symptoms of Pesticide Poisoning

The following table describes likely symptoms in cases of acute poisoning. This includes accidental spillage on hands or skin, swallowing or eye contact during the mixing and use of pesticides. It also includes exposure from aerial spraying or drift from a tractor or hand-carried applicator. The smaller doses, which may be had from touching or eating freshly sprayed crops or other objects, can produce similar symptoms. Much longer-term, or chronic, exposure to lower doses through residues is less likely to produce the symptoms listed here.

Very mild symptoms, especially those from very low doses in residues, are usually very difficult to link with pesticides. Often these symptoms include headaches, dizziness, irritability, mild asthma, or rashes, all of which are among the usual symptoms from a wide range of common diseases. However, even if you only *think* the symptoms could be due to residues, mention it to the doctor and try to remember where the residue could have been encountered.

Any suspected poisoning should be reported to a doctor or local hospital casualty department as soon as possible. Administration of any antidote and treatment for any effects of the chemical should be prompt. Always have available for the doctor the brand name of the chemical and the full ingredients list from the product label. Take the label if possible. This will help identify the best treatment.

The first-aid suggestions listed in the table are no substitute for medical treatment. They may, however, save a life or reduce any lasting damage in serious poisonings.

General advice

The National Poisons Information Service gives the following general first-aid advice:

1. If the person is unconscious, make sure they are breathing and resuscitate if necessary.

2. If the person is drowsy or unconscious but breathing, place in the recovery position.

3. If the person is alert, keep them calm and avoid all stimulation. Watch their breathing for any difficulties and for their state of consciousness.

4. Remove any contaminated clothing and wash off any chemicals with soap and water.

5. If the person is unconscious do not try to make them vomit and give no liquids by mouth. If they are awake, give water to drink, not alcohol or milk. Encourage vomiting immediately only if the pesticide is known and the table below advises urgent vomiting. Otherwise get medical help from a doctor or local hospital quickly.

Pesticide group	Symptoms	First aid
Organochlorines including DDT, aldrin, dieldrin, lindane, endrin, endosulfan	Skin contact or swallowing are the common routes of entry. The chemicals work by disrupting the nervous system and produce twitching and tremor which can lead to convulsions. Person can become apprehensive and excited within a few hours of exposure.	Remove contaminated clothing and wash skin in soap and water. Take person to hospital quickly for stomach emptying if pesticide has been swallowed. Do not give any fatty drinks or food (e.g. milk).
Methyl bromide fumigants	This is a gas which is absorbed by breathing. Acts as a cell toxin. Symptoms include skin burns from the liquid, leaving itchy blisters. Long-term use can leave skin rashes and cracking skin. Lung damage follows from breathing the gas, along with damage to the eyes. There may be headaches, irritated	Skin contamination requires removal of clothing and washing. Breathing of the gas means urgent hospitalisation is necessary. Keep person still and calm, and watch for signs of difficulty in breathing.

Pesticide group	Symptoms	First aid
	eyes and mouth, and abdominal pain. Damage to the nervous system can cause headaches, blurring of vision, loss of balance, and tingling in hands and feet.	
Carbamates of the insecticide type	These are anticholinesterase chemicals, and interfere with the sending of nerve signals. They enter the body by swallowing or by breathing in spray. Symptoms – headache, nausea, difficulty with breathing – appear more rapidly than with the organophosphates. There may be constricted pupils and muscle twitching with stomach cramps and diarrhoea. In severe cases there may be heart and breathing failure causing death. A chronic exposure to even small amounts could lead to a build-up of chemicals, with progressive damage to nerves which may appear only some time after exposure.	Treat as an emergency for hospital admission. Meanwhile, remove contaminated clothing, wash skin with soap and water, and keep person as still and calm as possible. Moving about or any action which increases the breathing rate could increase the severity of symptoms.
Carbamates, especially the **dithiocarbamates** (EDBCs) including maneb, thiram, and zineb	These are relatively safe, and the main route of absorption is swallowing. Large doses may cause irritation of mucosal surfaces in the mouth	Vomiting should be induced quickly, as there is no antidote once the chemical enters the blood stream. Because of possible problems

Pesticide group	Symptoms	First aid
	and throat followed by vomiting. There may also be difficulties in breathing and even paralysis.	with breathing, the person should be taken to hospital quickly.
Diquat and paraquat	Sometimes entry is via skin, but most often it is by mouth. These chemicals are directly toxic to specific organs like lungs, kidneys, or liver, which are reached by the blood after absorption in the stomach. Skin contact causes irritation and rashes. Rashes and blisters can form around the mouth and eyes if contaminated. Symptoms on swallowing include vomiting, soreness of throat and mouth, and diarrhoea. Large doses can kill within 48 hours from organ failure. Smaller doses may be followed by kidney, liver or lung damage within two to four days. Paraquat can cause fibrosis of the lungs, which reduces their efficiency and can cause death.	As little as one dessertspoon can kill. If swallowing is suspected, make person vomit immediately. Early admission to hospital is important, but if not possible, give doses of epsom salts and plenty of water. The extent of damage and prospects of recovery depend on the speed of emptying the stomach, as there is no antidote.
Organophosphates	These are anticholinesterase chemicals which disrupt the sending of nerve signals along nerve fibres. Absorption is by breathing or skin contact. Symptoms	Remove contaminated clothing and wash skin with soap and water. Keep person still and watch out for breathing difficulties. He or she should be taken to hospital as an emergency.

Pesticide group	Symptoms	First aid
	include exhaustion, fatigue, and mental confusion, followed by vomiting, stomach cramps, and sweating. There may be muscle twitching later in the eyes and tongue followed by the face, and then any other muscles in the body. There can also be diarrhoea, and breathing difficulties which may cause death.	
Chlorinated phenoxy compounds, including 2,4,5-T; 2,4-D; MCPA; and MCPB	Absorption is usually through the skin or from breathing in spray drift. Acute doses irritate the mucous membranes of mouth, lungs, and stomach, causing vomiting and diarrhoea. There may also be confusion, convulsions, and depression, with muscle spasms and weakness. Chronic effects include paralysis with 2,4-D and chloracne from the dioxins in 2,4,5-T.	Remove any contaminated clothing and wash with soap and water. There is no known antidote, so emptying the stomach is important. Induce vomiting or get the person to hospital quickly.
Includes chemicals which disrupt energy transfer in cells. Herbicides like **benzonitrile** chemicals, bromoxynil, and ioxynil, plus the **phenolics** like pentachlorophenol	Absorption is usually through the skin, but spray can also be inhaled. These chemicals interfere with energy transfer in the body's cells. Symptoms include increased sweating, thirst, tiredness, insomnia, and weight	Hospitalisation is essential, but first remove any contaminated clothing and wash skin with soap and water. Keep person calm, cool (with sponging if necessary), and stationary to avoid any unnecessary energy

Pesticide group	Symptoms	First aid
	loss. Long-term exposure (as with workers applying these herbicides regularly) can produce anxiety, rapid and deep breathing, rapid pulse, and high temperatures. Death can occur by total exhaustion.	expenditure.
Pyrethrins, natural and synthetic	Most easily absorbed through the skin and lungs. Most have very low toxicity to mammals, and are not likely to produce acute human toxicity, but some, like decamethrin, are highly toxic. They work by interfering with the sending of nerve signals. Symptoms may include twitching, excitability, and convulsions which can cause death if breathing and the heart fail.	Vomiting should be induced, preferably in a hospital, if the chemicals have been swallowed.
Organomercurials – a varied group, usually in antifungal treatments for seed and bulbs. Includes phenyl mercury acetate (Ceresol) and methoxy mercury acetate (Panogen) May be in mixtures with chemicals like lindane (Mergamma).	This varied group of chemicals has a range of toxicities but the straight-chain type disrupt the brain, possibly through interfering with enzymes. Absorption is through the skin or by breathing spray. Skin contact can burn, with blisters and redness. Symptoms from spray drift include irritated eyes,	Contaminated clothes should be removed and skin washed with soap and water. If pesticide has been swallowed, the person should be taken to hospital to have the stomach emptied.

Pesticide group	Symptoms	First aid
	nose, throat, and mouth, and can cause stomach upsets and kidney damage. Once absorbed into the blood, symptoms of tiredness, concentration and memory loss, and numbness and tingling of the lips, fingers, and feet occur. Serious poisoning may cause total loss of balance, disturbed heart rhythms, and tunnel vision.	
Inorganic mercurials include the fungicides mercurous chloride (Calomel) and mercurous oxide, which is used for canker control in apple trees.	Usually absorbed by swallowing. The oxide irritates and corrodes cells on contact. Produces a metallic taste followed by thirst, stomach pain, vomiting, and diarrhoea, often with blood streaks. If it gets into the lungs, failure of breathing may follow. The chloride is less toxic.	Small amounts of water should be given to dilute the chemical in the mouth and gut while person is taken to a hospital for stomach emptying and treatment.
Arsenics used to be found in some rat and mice poisons and in timber treatments, but are no longer approved for these uses. They may still be in old products lying about in sheds and warehouses.	These are highly toxic and enter the blood most rapidly through the stomach after swallowing, but can also get in through skin or lungs. They poison by causing dilation and leakage of blood vessels and also disrupt cell metabolism by blocking enzymes. Symptoms are rapid stomach pain and	Same as for **organic mercurials**.

Pesticide group	Symptoms	First aid
	vomiting, diarrhoea, and dehydration. Low blood pressure, muscle cramps, and severe electrolyte loss (e.g. of calcium and potassium ions) which can cause death. Chronic exposure to lower doses can have a cumulative effect, causing loss of appetite, diarrhoea, and the specific 'rain-drop' skin rash.	
Nicotine	Absorption is through skin, by breathing in spray, or from swallowing. The effects are due to toxicity to nerves. Symptoms include nausea, vomiting, and dizziness. There may also be headache, sweating, salivation, rapid pulse, and possibly coma or death within minutes.	Doses less than 1 g can kill and there is no known antidote. If nicotine is swallowed, induce vomiting quickly and get person to a hospital. Remove all contaminated clothing and wash in soap and water.
Sodium chlorate, a total herbicide found in household products like path and drive weedkillers	Absorption is usually through the stomach after swallowing. They are strong oxidising agents which damage the blood cells, stopping them carrying vital oxygen and damaging the kidneys. Symptoms include abdominal pain, vomiting, blood cell destruction, cyanosis, kidney damage. Death may follow a dose of a dessertspoonful.	Immediate vomiting is important. Hospitalisation is necessary because blood transfusion and dialysis may be necessary.

Pesticide group	Symptoms	First aid
Anticoagulant rat and mice poison, including warfarin	Absorption is by swallowing the baits. Toxicity is through interfering with blood clotting. Symptoms include bloody urine and stools, easy bruising, and delayed clotting of wounds.	Human poisonings are not often serious, but children may eat baits. Take person to doctor or hospital for treatment.

(Source: Based on *Pesticide Poisoning; notes for the guidance of general practitioners* (1983) London: HMSO.)

Appendix A: How Pesticides Work

The following are descriptions of the most common ways in which pesticides interfere with life processes.

Blocking photosynthesis

In photosynthesis, the creation of basic biological energy occurs. All life on earth depends on this process. Put simply, carbon dioxide and water are converted into sugars which are available as an energy source. In the process, oxygen is released, which is one reason why preserving the world's vast rainforests is crucial. The chemical formula for this reaction is:

$$6CO_2 + 6H_2O + \text{Light energy} \rightarrow C_6H_{12}O_6 + 6O_2.$$

There are two steps in photosynthesis. The first is capturing light energy and converting it into chemical energy. The energy from sunlight is absorbed in the green chloroplasts of leaves and used to split water, releasing oxygen and energy. This energy becomes attached to a special energy-transfer chemical found throughout living things and called NADP. Its chemical name is nicotinamide adenine dinucleotide phosphate.

The second step is the transfer of the energy attached temporarily to NADP to a more versatile storage form. This is in a high-energy phosphate bond in a substance called adenosine triphosphate, or ATP. These bonds take a large amount of energy to create and, once formed, can be stored or shipped around the cell or other parts of the organism to where energy is needed to power biological reactions. They work like rechargeable batteries, which can be used to power a wide collection of useful gadgets, from torches and food-blenders to computers. ATP is one of the most important substances in living organisms. Photosynthesis creates portable ATP energy from the energy of light.

This light energy, now in the combined form of $NADPH_2$ and ATP, is used to convert the carbon in carbon dioxide into carbohydrate carbon as glucose ($C_6H_{12}O_6$). This sugar is water soluble and can be moved around inside the plant from the leaves to any other part where it is needed, or it can be converted to insoluble starch for storage (roots, tubers, fruits). Any pesticide that disrupts the energy-transfer system thus blocks the process of photosynthesis and halts the plant's vital, and only, source of energy for growth.

Interfering with energy release

For a cell or organism to move, grow, form reproductive cells, or heal injuries, it needs a source of energy to perform the metabolic processes involved. Plants or algae that photosynthesise use the sugars, or their storage form as starch, for energy. Some plants also create stores of fat that they can mobilise for energy. (Our kitchen vegetable oils come from the energy stores of fruits or seeds – soya, sunflower, and olive oils.) Insects, fungi that thrive on living or dead plant material, and higher animals use sugars, fats, or proteins for their energy. Even with carnivorous animals, or omnivores like humans, all biological sources of energy can be traced back to photosynthesis. We can use the protein and fat in beefsteak for energy, but the free-range cow gets energy from the sugars and starch in the grasses and other plants that make her diet. The basic processes for releasing energy are the same in all living things.

Sugars like glucose release their energy in two steps. The first, called glycolysis, splits the sugar into two smaller molecules. This requires some energy which is provided by the same ATP energy store created during photosynthesis. Several times as much ATP is produced in the next stage of releasing the energy from the sugar. Glycolysis can occur without oxygen, which is what happens when sudden bursts of muscle energy are needed in running a marathon race. The sugar is incompletely 'burnt', with lactic acid as the end product. Another example is the fermentation of grape sugars in wine or added sugar in beer and lager, producing alcohol as the end product. The fact that alcohol still contains energy from the original sugar is demonstrated by the fact that it will burn, giving off heat.

In the presence of oxygen, the second step breaks the smaller molecules right down to carbon dioxide (CO_2) and water, with the release of more energy. This happens in a complicated cycle of reactions known as the citric-acid or Krebs cycle. In this cycle a single molecule of sugar like glucose is broken down, and 38 molecules of ATP are produced. This mass of ATP is used in the thousands of chemical reactions which keep organisms growing, moving, reproducing, and repairing damaged tissue.

Fats and oils are broken down to their component fatty acids that can be 'burnt' in a similar way to release energy. Once broken down to smaller molecules, the 'burning' of fatty acids follows the same citric-acid cycle as sugars. Proteins can also be used for energy as a last resort when fats or sugars are not available. This clearly is a last resort because, unlike sugars and fats, the principal role of which is energy storage, protein is a building material of structural tissues like muscle, tendon, and skin. When protein is 'burnt', it is first broken into the individual units, or amino acids, which make up the large protein molecules, and these are fed eventually into the same cycles that sugar and fat go through. Both fat and protein 'burning' produce ATP.

These processes in living organisms are more complex and varied than are pictured here, but the general picture is true. In each energy-release mechanism, whatever the source of energy used, are numerous interconnecting steps, many of which are controlled by specific chemicals called enzymes. Any pesticide which disrupts these cycles can reduce or stop the release of energy from stores. Without energy, all normal functions of cells and organisms stop, and death follows.

Blocking nerve signals

Hormones travelling in blood, or nerve signals channelled through nerve fibres, control every action of our bodies, whether conscious or unconscious. The nervous system allows the detection of stimuli like light and pinpricks, and controls reactions like jumping up after sitting on a drawing pin. Conscious activity, like turning the pages of this book, or unconscious activity, like the constant beating of the heart, are also controlled by nerve signals.

The nervous systems of insects and mammals have much in common and all work on the same principle. They consist of nerve cells, or neurones, which have a body and a long tail called the axon. The axons are like the wiring inside a house or pocket radio; they carry the electricity from the mains box to every room and appliance.

Nerve signals are small electric currents that travel along the axons from neurone to neurone and can move at about 270 mph. You may not be surprised to learn that ATP is again necessary to power the creation of the electric current.

When two neurones meet and the signal has to pass from one to the other, a connection has to be made, just as in electrical circuits. This connection is called a synapse. The same need for a connector arises when an axon reaches a muscle it controls. The connector is a chemical. There is a small space between two connecting axons or between a terminal axon and a muscle cell. When the electrical signal reaches the end of an axon, it stimulates release of a chemical transmitter into the space. This is most often a chemical called acetylcholine. The acetylcholine 'stimulates' an electrical charge on the second axon (or the muscle fibre), and the electrical wave continues its travel along the next neurone or into the muscle fibre. Once the chemical transmitter has passed its signal across, it is inactivated by a special enzyme called cholinesterase. If it remained in the space, it would block any response to any further signals.

Pesticides can work at two points to disrupt the sending of nerve signals. First, they can stop the creation and movement of the nerve signal. Second, they can destroy the cholinesterase enzyme, which prevents nerve signals going to or from the brain, and stops them connecting with the muscles they control. The result can be paralysis or erratic and incoherent nerve signals. Death usually follows. Several groups of pesticides are known to work through anticholinesterase activity, e.g. carbamates and organophosphates.

Damaging the central computer: DNA and RNA

In the centre of almost all living cells (mature red blood cells are one exception) is a nucleus that contains the chromosomes carrying the genetic material, or genes, of the organism. Chromosomes are really the central computer of living things. The chromosomes are made of DNA, or deoxyribonucleic acid, and genes are segments on chromosomes. In humans there are 46 chromosomes, 22 matched pairs and two sex chromosomes (X and Y). Women have a pair of Xs and men have an X and a Y.

Chromosomes are duplicated in cell division. In growing tissue, cells divide repeatedly, making exact copies of each other and increasing the number of cells and the total size of the growing organism. This division is called mitosis. When sex cells, sperm and ova, or eggs, are made, a different division called meiosis occurs. This results in only half the parent cell's number of chromosomes in each sperm or egg cell. At fertilisation, the chromosomes in the sperm and egg combine, making up the original number of pairs in the parent.

In cell division and in the copying of the chromosome material, there is plenty of opportunity for 'mistakes' to occur. These are called mutations. Most are disasters for the organism, and cells with mutated chromosomes usually die. Only a minute number of changes leave an organism better adapted to the world. Some mutations occur as errors in the complex copying procedures, but others are due to the damaging effects of radiation or of chemicals in the environment. There are some pesticides that aim to interfere with the duplication of DNA, and some that accidentally result in serious changes to the chromosomes. Both are known as mutagens (causing mutations) or teratogens (causing malformations to the developing foetus). Carcinogens are similar chemicals in that they also disrupt genetic material and cause cells to grow rampantly, without the normal guiding and controlling mechanisms written into healthy genes.

The second important part of the computer is RNA, or ribonucleic acid. This is similar to DNA in structure. RNA is concerned with the manufacture of proteins in cells. Proteins exist in a very large number of shapes and sizes, with functions more varied than simply providing the bulk of muscle. The enzymes and hormones which regulate most body processes and reactions are proteins.

There are several pesticides, especially among the fungicides, which work by disrupting protein synthesis. The synthesis of particular proteins or the blocking of RNA, which stops all protein synthesis, may occur. This usually disrupts vital enzymes, blocking normal functioning, and death follows.

Interfering with growth regulators

Plants, insects, and higher vertebrates like humans all have their growth regulated by hormones. In plants, hormones like indoleacetic acid and the gibberellins regulate cell growth. They ensure that leaves and stems grow towards the life-essential sunlight, as well as ensuring that the rapid growth of new cells in growing shoots and roots maintains strict order. There are also hormones which regulate the time of flowering and leaf-drop in the autumn. Insect hormones control growth in larval stages, the moulting, and the formation of the waxy, protective cuticle, or shell, of adult insects, as well as the development of sexual organs. Human hormones of comparable importance are thyroxine, which controls the pattern and rate of growth, and the sex hormones, testosterone, progesterone, and oestrogen. Such hormones control, among other things, the masculine body appearance and the preparation of the womb for nurturing the growing foetus.

Chemicals which mimic or inhibit such hormones can block growth or reproduction in pests. Some herbicides work on the mimic principle, and there is increasing interest in developing more hormone-mimic insecticides rather than poisons which can have serious environmental effects. Hormone mechanisms offer possibilities for targeting of pesticides very specifically, as well as for greater environmental safety.

Appendix B: Chemical Families of Pesticides

The following are the larger families into which the most commonly used pesticides fall. Some chemicals do not belong to any of these families, but they are the exception.

Organochlorines

These are based on a benzene ring with one or more chlorine atoms attached, either to the ring or to side chains. The best-known member is DDT.

DDT Related compounds were developed rapidly as the insecticidal power of DDT became legendary, but few attempts produced insecticides as powerful as the parent, DDT. Examples include methoxychlor and DDD. DDT is now banned in the UK, but it still appears in food residue tests at levels suggesting that it is being applied illegally on some crops of lettuce, mushrooms, and tomatoes.[1]

Hexachlorocyclohexane (HCH) or lindane (BHC) More soluble than DDT, lindane, otherwise known as BHC, is widely used in insecticide mixtures in the UK. It is used in soluble soil treatments as well as for fumigating or spraying against a wide range of insects. Lindane is under review by MAFF. Common products include Boots Ant Destroyer, Murphy's Gamma BHC Dust, and shampoos and creams against human lice and scabies.

Cyclodienes – the 'drins' Chlordane and heptachlor were early entrants to this group, but aldrin and dieldrin are the best-known members and are named after the two scientists who discovered the creating reaction – Diels and Alder. Heptachlor has a minor use in the UK in worm-controlling chemicals like Mascot Wormort and Parker's Wormex. The cyclodienes are usually complex, three-dimensional combinations of fused pentane, or five-carbon rings.

Aldrin, endrin, and dieldrin have similar structures and properties. Like DDT, they are soluble in fats and do not degrade easily in the environment. They are effective insecticides, working by contact, and hence are used as a soil insecticide or are applied to surfaces visited by flies and ticks and on fabric where moths may lay eggs. Total bans on the use of aldrin and dieldrin in the UK have been in force since 1989, while endrin has been banned for much longer. Because of their persistence in the environment, their residues will continue to be found in the food chain for some years to come. Despite being an established carcinogen and highly persistent, chlordane is still permitted in the UK for use on public-amenity turf against worms.

Other organochlorines Chemicals like pentachlorophenol, which is similar to lindane, are used in herbicides, fungicides, and wood preservatives. Several important fungicides like captan and captafol are based on the benzene ring but contain sulphur as well as several chlorine atoms. The fungicide quintozene is also similar to lindane but has nitrous oxide in place of one of the chlorine atoms on the ring.

Mode of action Despite being among the earliest of mass-produced pesticides, it is remarkable and humbling that we still know very little about how organochlorines act. Clear hints are given in the writhing, jerking death throes of insects under their influence. It seems they interfere with the transmission of nerve impulses, either eliminating or jumbling the signals. Without clear nerve signals such vital functions as breathing cannot occur, and general body muscles twitch erratically and uselessly. How the sulphur-containing fungicides in the group work is unknown, but sulphur has long been known to be toxic to fungi.

Health and environmental effects Organochlorines vary in their acute toxicity for humans from an LD_{50} (see page 72 for definition) in rats of 3 mg/kg for endrin to 457 mg/kg for DDT. A few drops of endrin or a tablespoon of DDT may kill an adult human! Acute poisoning causes dizziness, nausea, and twitching of muscles in arms and legs, followed by tremors and convulsions. Breathing may speed up at first but can stop completely. Chronic poisoning may arise because of the tendency of the group to accumulate in fatty tissue within the body, causing kidney and liver damage. Long-term build-up in body fat can also lead to sudden release and symptoms of acute poisoning if fat reserves are mobilised, for example, with sudden depletion of excess fat during dieting or a period of psychological stress.[2]

This group has also been found to contain carcinogens, mutagens, and teratogens, which is one of the main reasons they are banned, or at least severely restricted, in many countries.[3] DDT, aldrin, dieldrin, and heptachlor are recognised carcinogens and teratogens. Lindane is the only one of the group approved for use in the EC, but it is under suspicion as a carcinogen and mutagen. It is suspected of causing leukaemia among the factory workers who make it.[4]

Organochlorines are broad-spectrum insecticides, meaning they kill a large range of insects which may include harmless or beneficial insects as well as the target pest. They also harm fish, birds, and mammals.

This entire family is distinguished by its persistence in the environment. The chemicals are not easily broken down in soil or water so they accumulate. DDT has a half-life of 20 years in soil. One headache with long-term persistence is the speed with which insects develop resistance. Organochlorines also have the awkward habit of accumulating in the fatty tissue of insects, birds, fish, and animals. This causes the food-chain multiplier effect, in which doses too small to kill organisms accumulate in body fat and become concentrated as animals higher up the food chain eat large numbers of these lower organisms as their food. DDT travels up the chain through algae and plankton, insect larvae, small fish, and rodents, through birds and larger mammals to still larger birds and animals, concentrating at every step. Rare birds of prey still die each year in the UK because of this effect. In February 1990, two of 11 reintroduced red kites were found dead from endrin poisoning even though endrin was banned a decade ago.

Halogenated hydrocarbons

Unlike the organochlorines, this group is based on straight chains of carbon atoms to which are attached not only chlorines but also other halogen atoms like bromine and fluorine. The group includes herbicides and nematicides as well as insecticides. They are often used as fumigants because they vaporise easily and can thus diffuse through stored grain, glasshouses, and the soil.

Bromide group   Among the early and effective fumigants are methyl bromide and ethylene dibromide. These are toxic to mammals, and methyl bromide, in particular, is extremely toxic to humans. Now banned in Europe as a means of killing insects in stored fruits and vegetables, the toxicity of ethylene dibromide has been used as a justification for food irradiation as an alternative pest control.

Dibromochloropropane (DBCP) is a largely phased out nematicide fumigant used in the 1950s on Third World crops of tobacco and pineapples.

Chlorinated group   Two pesticides against nematodes come from propane and propene gases with the addition of chlorine – dichloropropane and dichloropropene. Propionic acid is the starting point for two useful herbicides, especially active against grasses. Dalapon is a translocated herbicide useful against grasses like couch, and its cousin, trichloroacetic (TCA) acid, also hits grasses but is less effective. It works in the soil rather than by travelling up inside the weeds.

Mode of action
The nematicides in this group attack nematodes by being lipid (fat) soluble and penetrating the outer protective coat of these worm-like organisms which live in soil. Once inside, they act like narcotics and depress the activity of the nervous system. The herbicides disrupt and block normal activity in the plant cells they contact by precipitating proteins and therefore deactivating the important cellular enzymes

which control metabolic reactions. Pantothenic acid, an important B vitamin in both human and plant nutrition, is one enzyme destroyed. Damage can be widespread.

Health and environmental effects
The volatility of halogenated hydrocarbons and their solubility in fact make their entry into human bodies relatively easy, and they become trapped in fat tissue. Like chloroform, which is nothing other than trichloromethane ($CHCl_3$), these chemicals anaesthetise the central nervous system. One old pesticide (chlorpictin) was made from chloroform. The chlorine and bromine atoms are particularly toxic to the peripheral nervous tissues. Kidney damage can occur in chronic exposure. They are active alkylating substances, which means they can damage DNA, causing mutations and cancers. Ethylene dibromide is classified as a potent carcinogen and mutagen with the ability to penetrate protective clothing, rubber, and, with ease, human skin. DBCP has been proved to be a sterilant of male factory workers and a carcinogen.

While the herbicides are reasonably specific to such grasses as couch and wild oats, the soil fumigants have a broad-spectrum effect, killing off many harmless and even beneficial organisms in the soil. Some, like dalapon, are also very mobile within the soil and can be washed into ground water. Wide use of DBCP on cotton in the US forced the closure of underground wells.

Carbamates

This is an interesting group which developed from the work on organophosphorus insecticides which also act by disrupting nerve signal transmission. They are derived from carbamic acid, which does not exist in the free state.

The Geigy Company introduced the first effective insecticidal member of this group – Isolan – in 1951. It is water soluble and effective against aphids and houseflies, but it is so toxic to mammals that it was withdrawn. New carbamates based on a pyrimidine ring, that is, a benzene ring with two nitrogen atoms replacing two of the carbon atoms, were found to be safer. Pirimicarb was less toxic but a good systemic aphicide. Later, phenolic rings were used, producing carbaryl, also known as sevin. Carbaryl is a broad-spectrum insecticide used on vegetables, fruit, cotton, and turf. It was an early replacement for DDT because it breaks down and leaves no harmful residues.

Other insecticidal carbamates include cabofuran, propoxur, aldicarb, and methomyl. One chlorinated carbamate which was developed as a pre-emergence herbicide is chloropham, and a relation, barban, was used on established weeds.

Thiocarbamates Sulphur-containing carbamates, known as thiocarbamates, like Eptam (EPTC), are pre-emergence herbicides used against grasses and broadleaf weeds like fat hen and chickweed in potato crops.

Out of the vulcanised rubber industry emerged the first pure organic fungicides – the dithiocarbamates. The first of the bunch was thiram, which is still used. The most important family within the group are the ethylenebisdithiocarbamates (EBDCs), which are widely used fungicides. The first, nabam, is converted into its siblings zineb and maneb by reaction with zinc and manganese salts, respectively. The EBDCs were the subject of an urgent review by the government during 1989 after they were dropped from the US lists on grounds of carcinogenicity. They remain on the approved list in the UK. (See pages 66–7.) Ring-structured carbamates were developed with varying specificity for grasses or specific broadleaf weeds like dock and bracken. Tri-allate was a sulphur-containing carbamate aimed at wild oats. Asulam appeared in 1965 as a successful treatment for the bracken which is still spreading over the upland hills of the UK.

Mode of action

The carbamate insecticides are anticholinesterase in action (see Appendix A) and disrupt the nervous system. This means disordered nerve signals, and, where muscles are concerned, there are convulsive jerkings and eventual death from respiratory or heart failure.

The fungicides may work through several different actions. The EBDCs seem to break down readily into toxic substances that react with important thiols compounds in the fungal cells, or they may bond strongly with trace elements like copper in the cells, starving the cells of the use of the copper. Other members block energy transfer in the Krebs cycle, which is crucial in the metabolism of cells. (See Appendix A.)

Some of the herbicides work by blocking the photosynthetic reaction. (See page 295.) The absence of photosynthesis causes death as soon as stocks of carbohydrate run out. Others disrupt the process of cell division, mitosis. (See page 298.)

Health and environmental effects

Aldicarb (Temik) is one of the most toxic pesticides available, with an oral LD_{50} of 0.93 mg/kg for rats and an ADI for humans of only 0.005 mg/kg. It is also toxic to wildlife and fish. The group is water soluble and easily absorbed through the skin or on breathing. The signs of serious poisoning are typical of anticholinesterase pesticides, and chronic effects include depression, nerve damage, and poor memory. They break down into ethylene thiourea (ETU) in sunlight, and in food processing and cooking. ETU is an animal carcinogen, producing thyroid cancers. Although the EBDCs are not absorbed into the sprayed plants, commonly potatoes and wheat in the UK, the breakdown products, or ETUs, are absorbed. Residues are found in lettuce, spinach, and fruit like apples and pears.

With the exception of aldicarb, the group breaks down readily in the environment. They can affect such non-target organisms as fish and earthworms.

Heterocyclic compounds

This is a large group of successful herbicides based on a ring structure containing both the usual carbon atoms and nitrogen atoms. Many are so effective that they are used as total herbicides on sites like car parks or around buildings, where they destroy everything green and growing. The same herbicides in smaller doses are more subtle, killing only broad-leaved weeds when used on crops like sugar cane and asparagus or around soft-fruit bushes like gooseberries. Grass family crops, including sugar cane and maize, are unaffected by members of the group, like atrazine, because they contain special enzymes which break down the herbicide as soon as they get into the plants.

Triazines   These are soil-acting herbicides that are taken into the weeds through the roots and kill by destroying the photosynthetic function of the leaves. The best-known chemicals are atrazine and simazine, which were developed in the 1950s following the discovery of their herbicidal properties by the Geigy Company in Switzerland. Related chemicals with slight modifications to the original structure include prometryn, desmetryne, terbuthylazine, and cyanazine. These are important pre-emergence herbicides in field crops. They are sprayed onto the soil and kill weeds as they emerge through the soil into the light.

Triazoles   Amitrole was introduced as a herbicide in 1954. It is a non-selective herbicide used to keep fallow land clear, and against perennial weeds between trees in orchards.

Bipyridyliums   These important herbicides have played their part in the farming revolution. Functioning only in sunlight, they desiccate soft tissues such as leaves and so kill the plants on to which they are sprayed. Total herbicides, they have made a dramatic impact on farming since they were developed in the 1960s. Farmers plough fields to kill weeds and prepare a seed bed. But having tractors haul heavy machinery many times across fields costs money and damages the soil. So these plant desiccants were a clever answer to several farming problems: they could be sprayed on an unploughed field, killing every growing weed. Not only was ploughing then unnecessary, but also the dried weeds formed a protective layer on the surface of the soil, reducing erosion and providing a soft and protective surface mulch into which the seeds for the next crop could be gently inserted by machine. This was the advent of chemical ploughing.

Another advantage of bipyridyliums is that when they fall on the land rather than the plants, the chemicals are immediately and tightly bound to the soil. Thus they cannot be taken up by the roots, but, most important in chemical ploughing, they cannot affect the planted seeds when they germinate.

The two most often used herbicides in this group are paraquat and diquat. They are based on combining two pyridine rings, which gives them their group name bipyridyl.

Other heterocyclics   Relatives include the substituted uracils like bromacil, which contains bromine in the molecule; terbacil with chlorine; and the thiadiazine chemical, bentazon. With the exception of bentazon, which acts on the leaf surfaces it contacts, these are also soil-acting herbicides.

Mode of action

In one way or another, all the heterocyclics disrupt the photosynthetic activity of leaves, starving plants of their energy. The triazines block important reactions creating ATP and NADP within the photosynthetic process. (See Appendix A.) The triazoles block the plant's manufacture of the carotenoid pigments which protect the vital chlorophyll pigments in the leaves from destruction by sunlight. Dying leaves turn yellow as their green chlorophyll pigments disappear. The bipyridyliums work slightly differently by a sinister method. They need both light and oxygen to work. Oxygen is released in the chloroplasts during the photosynthetic reactions. The oxygen reacts with the paraquat or diquat, producing hydrogen peroxide, which destroys the membranes around the vital chloroplasts and other working parts within the plant cells.

Health and environmental effects

Most of the group have low acute toxicity to humans but may cause skin or eye irritations. Over the long term there may be cancer risks from simazine and amitrole, and both simazine and atrazine can form cancer-causing nitrosoamines in the stomach.

The bipyridyls are more toxic than the triazines. They affect skin and mucosal surfaces of the nose, throat, and other respiratory passages. They cause ulceration and inflammation, and delay the healing of cuts and wounds. Diquat, on being swallowed, damages the gut, and paraquat tends to damage the lungs and kidneys. There is no known antidote to this poisoning. Both have produced deformed embryos in rats.

Atrazine and simazine are among the most frequently found pesticide contaminants of water supplies in the UK.[5] Atrazine may be leached through the soil into underground waters or be washed into streams and rivers, where it can last for several years. The Government has placed both on a special 'red list' of dangerous substances over which there should be tighter controls in water supplies.

Paraquat and diquat are strongly adherent to soil particles and break down rapidly on contact with the soil. However, they have been shown to kill blue-green algae in the soil, and paraquat, in particular, is toxic to fish and mammals, including hares. Users are advised to keep all livestock out of treated fields for at least 24 hours.

Organophosphates

When the environmental persistence of the organochlorines was realised, the less persistent organophosphates were seen as the new generation of safer chemicals. Their toxic pedigree came from research into nerve gases during the Second World War. That they worked well as insecticides was clear, but the early organophosphate chemicals like dimefox were also very toxic to humans and safer versions had to be developed before they could be widely used.

The group includes a large number of insecticides, which were the first organophosphates developed; herbicides; and fungicides. They all contain phosphorus and an organic structure that can be straight chain or ring. The group illustrates the great variety of structures among pesticides, with some members fitting equally into other families like the heterocyclics or pyridines. Thus, there are chlorinated organophosphates like fenchlorophos and also heterocyclic N-compounds like menazon.

Insecticides Among the group are such well-known members as parathion, which is no longer approved for use in the UK. It is a non-specific insecticide that has a high toxicity to mammals (an acute LD_{50} of 13 mg/kg for rats) and destroys non-target insects like bees, and birds.

The introduction of chlorine atoms in this group was found to reduce toxicity to mammals without affecting the insecticidal powers. A series of such formulations appeared and was found, like many of the organophosphates, to be effective against insects which had become resistant to DDT. Bromophos is a contact insecticide against mites as well as caterpillars. It contains both chlorine and bromine and could also be included among the halogenated hydrocarbons.

A series including sulphur atoms sprang from demeton-S-methyl and includes phorate, a very toxic insecticide used against aphids and flies in crops like carrots and parsnips, and malathion. Malathion is an important treatment for aphids, thrips, and mites on a range of vegetables and fruits. It has a very wide range of susceptible insects and a low toxicity to mammals (an acute LD_{50} of 2,800 mg/kg for rats). It is a common ingredient in garden greenfly sprays.

Dichlorvos vaporises easily. It is used in household fly strips and has been used for flea collars. Its relatives, chlorfenvinphos and tetrachlorvinphos, fenchlorophos and mevinphos, are also non-specific insecticides of varying toxicity, all of which act on contact in the insect's stomach. Dichlorvos and chlorfenvinphos are highly toxic (with acute

LD_{50} of 80 mg/kg and 15 mg/kg for rats, respectively) tempered to tetrachlorvinphos, which has an acute LD_{50} of 4000–5000 mg/kg.

Dimethoate is a systemic insecticide and acaricide used against mites, thrips, and aphids on food crops. It is more residual on sprayed crops than dichlorvos. Livestock should be kept out of sprayed fields for a week. Most sprayed crops cannot be harvested for 7 days, but watercress must be left for 40 days.

Another group of organophosphates is based on heterocyclic structures like the triazines and pyrimidines. Azimphos-methyl and diazinon are broad-spectrum insecticides. Chloropyrifos is a contact and stomach-acting insecticide with a very wide range of victims and a persistence of 2–3 months. Although these three chemicals are widely used in the home and garden, all are dangerous if swallowed and can irritate the skin.

Fungicides and herbicides Following the development of organophosphate insecticidal compounds, organophosphate fungicides were explored. Wepsyn was an early systemic fungicide based on the triazole structure. Systemic fungicides were a great advantage against fungi that penetrate and grow within the host plant's tissues. They could perfuse all tissues of the plant and kill any growing fungal threads which appeared. Pyrazophos, an organophosphate based on a heterocyclic nitrogen structure, is an interesting combination of systemic fungicide and insecticide.

A well-known and widely used herbicide is glyphosate, which is a broad-spectrum, translocated herbicide active against couch grass as well as other perennial weeds and annuals.

Mode of action

The mode of action of organophosphorus insecticides is the same anti-cholinesterase activity seen in the carbamates. (See page 304.) However, different insects and higher animals handle these pesticides differently in the body, sometimes converting them to harmless products. This is why certain pesticides can be highly toxic to a small group of insects but not to others, or to humans.

The fungicides appear to act by disrupting enzymes, especially those associated with building the cell walls.

Health and environmental effects

The immediate effects of poisoning are seen in the functioning of the entire nervous system. First signs may be weakness and tightness of the chest with constricted pupils. Twitching and general muscle weakness may begin, and end in convulsions and respiratory failure. Very mild poisoning can lead to symptoms often mistaken for a virus infection, with sweating, headache, gastrointestinal upset, and weakness in the limbs. Such poisoning is difficult to diagnose because it is usually difficult to identify a source of low-level contamination with pesticides if it comes in food, grass, or dust. Low-level, long-term contamination can lead to damage to the nervous system, which may appear as depression, loss of memory, and anxiety. The organophosphates are easily absorbed through the skin, and severe irritation may occur where pesticides make contact.

Most organophosphates do not persist long in the environment. Exceptions such as glyphosate and chlorfenvinphos are strongly bound to soil particles, where they break down over a year or more. The group is very toxic to fish, birds, earthworms, and bees.

Chlorinated phenoxy substances

All plants have hormones that regulate growth. Natural plant hormones include auxins, gibberellins, and kinins. The auxins are concentrated in the growing shoots and are responsible for the elongation of cells, the bending of stems towards the light, and the downward growth of roots. An excess of growth hormone can produce chaotic growth often seen as bulbous galls on leaves and stems of plants. This effect was turned into herbicide activity through the discovery of chemicals that mimic the auxins.

The first such herbicide was 2,4-dichlorophenoxyacetic acid, better known as 2,4-D. This had its greatest effect on broadleaf weeds, less on grasses, and least on woody shrubs and trees. This meant that 2,4-D and its family were effective as selective herbicides. A more widely used relative is MCPA (2-methyl-4-chlorophenoxyacetic acid), which is an ingredient in many commercial mixtures of herbicides. The controversial shrub and tree herbicide 2,4,5-T contains the deadly chemical dioxin as a manufacturing contaminant.

Other herbicides in the family include mecoprop and dichloroprop, which are derivatives of propionic acid. MCPB and 2,4-DB are closely related chemicals based on phenoxybutyric acid.

Mode of action
The simulated hormone action of the family causes a disordered growth of the growing shoots which destroys any natural order or pattern of growth, and the plant dies.

Health and environmental effects
These pesticides cause mild-to-severe skin problems, ranging from irritations to chloracne, which may be caused by the dioxin. The chemicals can irritate the lungs and gut, and affect the nervous system. Symptoms of nausea, vomiting, diarrhoea, and mental confusion are reported. There is evidence that 2,4,5-T is responsible for birth defects and cancers among agricultural workers, and it has been the target of a trades-union campaign to have it banned in the UK. Since the Vietnam War, in which both 2,4-D and 2,4,5-T were used to defoliate the forests, significant numbers of birth defects have been attributed to these herbicides.

The family is relatively harmless to wildlife, especially as the chemicals are broken down rapidly in the soil. The dioxin contaminant persists longer than the herbicide, and it can affect reproduction in mammals. Evidence of a longer life in water comes from a Friends of the Earth study of contaminated drinking water in England and Wales that found mecoprop, dichloroprop, MCPA, MCPB, 2,4-DB, 2,4-D, and 2,4,5-T in both underground and surface water sources at levels exceeding the EC's maximum acceptable concentrations (MACs).[6]

Amides and ureas

A large family of herbicides was developed out of urea-based, phenyl-ring-structured chemicals. Fenuron and diuron were early examples. More recent additions to the substantial list of field crop herbicides are monolinuron and isoproturon.

Related amine chemicals were developed, e.g. propanil and solan, which are not approved for use in the UK, and alachlor, which is. More recent versions are naphthylacetamide and pendimethalin. An example of the growing complexity of synthetic pesticides and the way the families become combined is the herbicide cyanazine. Its formal chemical name makes the point: 2-chloro-4-(1-cyano-1-methylamino)-6-ethylamino-1,3,5-triazine. It has chlorine, a heterocyclic ring with nitrogen atoms, amine groups, and a triazine structure.

Another related group within the family are the anilines that provide herbicides like trifluralin, which is widely used in herbicide mixes.

Mode of action
The ureas disrupt photosynthesis and starve the plant of energy. The addition of chlorine atoms increases this effect but also increases the persistence in the soil. The anilines block seed germination and root growth by disrupting mitosis, or cell division.

Health and environmental effects
Most are of moderate-to-low acute toxicity but, because of the nitrogen component in the urea and amide structures, they can produce cancer-causing residues in the body. They are persistent in the environment. Many are used as residual herbicides to kill weeds emerging after the crop is planted. They are dangerous to fish, diuron especially so, and to algae in rivers or lakes which they reach in runoff.

Benzonitriles

This family of herbicides is important in field-crop weed control in Europe, usually in mixtures with other herbicides. They are contact herbicides against broadleaf weeds. The two most popular members each contain a halogen atom – bromine or iodine – which is reflected in their names. Bromoxynil and ioxynil are popular in agriculture, and illustrating the tendency towards mixing active ingredients in commercial products are the 25 different mixtures using bromoxynil and the 28 using ioxynil approved in the UK. Many of the available mixtures contain both chemicals. Neither of them is allowed in household herbicides in the UK.

Related herbicides are dichlobenil and its cousin chlorothiamid, in which sulphur replaces nitrogen in the original nitrile structure.

Mode of action
They disrupt the energy storage and transfer system in plants, and they also disrupt photosynthesis.

Health and environmental effects
These chemicals are harmful if swallowed or if they contact the skin or eyes. They block cellular energy-transfer processes in humans and also produce symptoms of increased cellular metabolism evident as fatigue, sweating, insomnia, and, over time, loss of weight. There is no known effective treatment, but symptoms can be relieved.

Benzonitriles are designed to persist in soil and on treated plants. It is advised that livestock be kept out of treated fields for at least six weeks. They are also harmful to fish, which is important because they persist on the foliage of their target plants, which may find their way into nearby ponds and rivers.

Phenolic compounds

The basic phenolic structure is a benzene ring with one hydroxyl group (OH) attached. The phenol group of pesticides has members with additional atoms or molecules attached to the ring at one or more of the ring's carbon atoms. The group contains herbicides and fungicides.

It has long been known that phenols are useful against human infections caused by bacteria, fungi, and intestinal worms. Plant medicines like thyme and garlic owe their antiseptic properties to phenols, as does the common weed, tansy, which kills human infestations of threadworms. Even the biblical myrrh was used as a fumigant, thanks to phenols. Phenols were used as early fungicides in agriculture but were often toxic to the plants they were meant to protect. Such compounds were suitable for 'dead' plant materials like sawn timber. Creosote oil is often used to treat wooden fencing in the UK.

The toxicity of these chemicals is reflected in the fate of the major dinitrophenols under the UK approvals system. From creosote came 2,4-dinitro-o-creosol, or DNOC. It is a strong stomach poison against many insects, also killing their eggs, and also mites, but is also used as a contact herbicide against broadleaf weeds. Its approval for use in the UK was withdrawn at the end of 1989. Its close relatives, dinoseb and binapacryl, have had their approvals withdrawn since 1987. Dinocap is a powdery-mildew fungicide now in use but under review.

Sodium pentachlorophenoxide is a timber and wall fungicide and mossicide used in mushroom houses in the UK. It is related to pentachlorophenol, which was discussed under the section on organo-chlorines.

Mode of action

These substituted phenols uncouple oxidative phosphorylation, which is the means of storing and transferring energy in plant cells. The loss of the transfer ability means that the organelles within the cell soon run out of energy supply, and the cells stop functioning and die.

Health and environmental effects

These phenols are toxic to humans and other mammals. They are absorbed through the skin and gut, and can damage the liver, kidneys, and brain. In the environment they are very toxic to fish and algae. They can vaporise from treated woods and are readily leached into surface and ground water supplies.

Synthetic pyrethrum compounds (pyrethroids)

Pyrethrum, from the flowers of *Chrysanthemum cinerariaefolium*, is one of the oldest natural insecticides. It is also one of the few natural compounds still widely used, and is enjoying particular popularity today with the recent increase in environmental awareness. Pyrethrum has a rapid 'knockdown' effect but some insects may recover and fly on to annoy another day. It is usually mixed with other, usually synthetic, pesticides to kill the insects while they are stunned. It is very safe for other wildlife and has very low persistence. In fact, contact with light and air begins its breakdown. This means that it is a poor replacement for some of the less safe pesticides now used in agriculture since most farmers want a pesticide to remain active for as long as is necessary to protect a crop or animal.

The commercial pyrethrum insecticide is made of a collection of six main chemicals from the flowers, each of which contributes to the insecticidal activity. The group is known as pyrethrins and individually as pyrethrin I and II, cinerin I and II, and jasmolin I and II. However, the names *pyrethrum* or *pyrethrin* on the product label should not be taken as a mark of safety. Many products are mixtures including some of the more toxic pesticides like lindane.

The pyrethrins have been synthesised, and many modified formulae are available, including isomers (compounds with the same formula but having some atoms arranged in a different pattern). The modified forms are known as pyrethroids. They include allethrin and its isomer, bio-allethrin; resmethrin and its isomer, bioresmethrin; and permethrin.

Again, these compounds are often found in mixtures, sometimes with other pyrethroids or with pesticides like dichlorvos or lindane. The pyrethrins are often mixed with another chemical known as a synergist that acts to increase the activity of the pyrethrin. Piperonyl butoxide is often used, and it makes the mixture seven times as effective.

Residual pyrethroids have been developed, like cypermethrin and deltamethrin, that are less sensitive to environmental breakdown and are stronger contact and stomach insecticides than the earlier derivatives of pyrethrum.

Mode of action

The whole family acts by interfering with the transmission of nerve signals in both the brain and the nerves among muscles and other organs. Small doses cause repetitive and erratic nerve signals, but higher doses stop all transmissions of signals.

Health and environmental effects

The natural pyrethrum is relatively safe to mammals, including humans, but the synthetic pyrethroids have varying toxicity, and they should be treated with the same respect as other pesticides. The more recent members like deltamethrin are more toxic to humans and cause skin and eye irritations. Some members like permethrin and allethrin are seen as possible mutagens or carcinogens.

Because the pyrethroids are not selective, they kill non-target insects, and they are particularly dangerous to bees. The later, more residual versions are a particular problem. Deltamethrin, for example, persists for three to four weeks on sprayed crops. All pyrethroids are extremely toxic to fish.

REFERENCES

1. MAFF (1989). Report of the working party on pesticide residues: 1985–8. *Food surveillance paper no. 25.* London: HMSO.
2. O'Brien, M. H. (1984). *On the trail of a pesticide.* Eugene, Oregon: Northwest Coalition for Alternatives to Pesticides.
3. IARC (1973). *IARC monographs on the evaluation of the carcinogenic risk of chemicals to humans.* 5: Oct. Geneva: WHO.
4. NIOSH (1974). *Criteria for a recommended standard occupational exposure during the manufacture and formulation of pesticides.* US DHEW No. 78–174.
5. FoE (1988). An investigation of pesticide pollution in drinking water in England and Wales. A Friends of the Earth report on the breaches of the EEC Drinking Water Directive. London: FoE.
6. FoE. Ibid.

Index

Individual pesticides (active ingredients) are listed alphabetically in the table on pages 204–85. Those pesticide products mentioned in the text are included in the index but a more exhaustive listing of available pesticide products for ordinary consumers or professional use is included under the active ingredient name in the table. The active ingredients will be found listed on the product label.

immune system, 61
Industrial Pest Control Association, 69
Institute of Terrestrial Ecology, 83
insurance spraying, 36
Integrated Pest Management, 36, 167
International Biotest Laboratories, 84–5
Invicta, 35
ioxynil, 311
IPM *see* Integrated Pest Management
isoproturon, 310

kestrels, 37
knockdown effect, 312
Krebs Cycle, 296
Kwik Save, 146

ladybirds, 46
Law Reform Commission of Canada, 85
LC_{50}, 72
LD_{50}, 72–3
leaf spot, 44
leaf stripe, 44
leather industry, 31
lice shampoo, 19
lindane, 19, 61
Local Authorities, 69
  monitoring residues, 126–7; pesticide use, 28–9
Long Ashton Research Station, 35
loose smut, 44
Low, W. M., 146
low-input farming, 167

Maclean, David, 66
MACs *see* Maximum Admissible Concentrations
MAFF, 68, 69, 70, 71, 86, 94, 117, 129
  public confidence in safety testing by, 100; relationship with farmers, 68–135
malathion, origin, 12
maneb, 20
Marks & Spencer,
  labelling, 156; monitoring, 153; pesticide policy, 148

Maximum Admissible Concentrations, 39
Maximum Residue Levels, 41
  altered list, 121–2; enforceable, 121–3, 135; exceeded on lettuce, 130; food and industry, 148–52; good agricultural practice, 136; need for decrease in levels, 136; official doubts over, 121–5; origins, 99; tecnazene, 117
mines, 31
Minister of Agriculture, 66–7, 84, 89, 92, 94, 136
miscarriages, 57
mitochondria, 13
monitoring residues, 126–7
  agencies responsible, 126; food industry, 152–5; reporting limit, 132; role of WPPR; sampling level 131–2, 135; surveys, 128–9, Total Diet Study, 128–31
Monsanto, 85
MRL Regulations, 109–10
MRLs *see* Maximum Residue Levels
mutagen, 55–6,
  tests for, 75
mutation, 42
myxomatosis, 15

National Association of British & Irish Millers
  monitoring, 152; pesticide policy, 145
National Farmers' Union, 143
  pesticide policy, 144
National Food Survey, 129
National Institute for Occupational Safety and Health, 57–60
National Poisons Unit, 50–1
National Rivers Authority, 54
nematicide, 17
nerve gas, 21
nervous system
  damage to, 62, 297; functioning of, 14, 297; signals 14, 20, 22; synapse, 297
nicotine
  poisoning advice, 293
nitrosamines, 306
No Observable Effect Level, 74, 77, 119–22